How to Make Money in
the Home Inspection Business

Walter W. Stoeppelwerth
Larry Vickers

HomeTech Information Systems
5161 River Road, Bethesda MD 20816
1-800-638-8292

Table of Contents

Table of Contents

Table of Contents

Table of Contents

Table of Contents

Table of Contents

1. INTRODUCTION

Historical Background

A generation or two ago in this country, people were so hungry for a home of their own that they grabbed almost anything for sale on the street. It was the end of a major war, and money was available to fuel an unprecedented-pent up demand for single family homes. Then during the 1950s, millions of 1000-square-foot "crackerbox" houses were built. They were not luxurious by today's measure, but these were the starter homes for people whose American Dream promised them independence, freedom, and privacy in their own family "castle".

A number of major changes in our society began with the first postwar construction boom. We were becoming more mobile than ever, and the promise of progress would gradually change the way Americans viewed their family home. For the first time, we began "moving up" in massive numbers, trading homes every few years much in the same way we traded cars. We went through the flight to the suburbs and a new commuter lifestyle, and as the population aged, those small houses built just after the war changed hands over and over again. Now the baby boom, the generation born to the original owners of those homes, has matured and they are becoming the first-time home buyer market. In many instances they are renovating those original residences and repeating the family pattern established by their grandparents.

The result of that expansion of material wealth is seen today in a national housing inventory of over 95 million individual dwelling units. At least 70 million of those are old enough to require major repairs or renovation. The blush is off the rose in these units, and yet they will remain in use for at least another generation to come, perhaps longer. As they change ownership, it's time for a new layer of expertise to enter the scene: home inspectors who can advise prospective new owners on what they should expect in the way of short and long term maintenance and repairs.

While those houses may have been reasonably well constructed for their time, the march of technology led to development of certain materials which were thought to be "miracles", only to fall into disuse once their full impact was understood. Two examples are DDT and asbestos -- both hailed as breakthroughs when they were first developed and only later discovered to be major environmental dangers.

Now DDT is no longer allowed on the fields, and asbestos is a liability in the home. It must sooner or later be removed from buildings where it was originally used as a mineral wonder, as siding, roofing, insulation, fireproofing in appliances and heating plants. There are thousands of places in existing buildings where asbestos may appear and not be obvious to the owner or occupant. It takes an expert, and sometimes even a testing laboratory, to positively identify asbestos. Then it takes special and expensive methods to have the cancer-causing material removed from the many places it may be hidden. This all requires a new level of technical expertise both to identify and to solve the problems.

Additional environmental conditions have come to be recognized as possible disasters waiting to happen, including odorless radon gas seeping up from the subsurface, interior air pollution from various sources, groundwater pollution, lead-based paint in areas where children might be exposed. Then there was the short-lived popularity of aluminum wiring which proved in certain instances to be dangerous. The qualified inspector must know when to call for an outside expert evaluation, and what signs to look for, to spot environmental as well as structural problems. In addition to these factors, it is important to know which safety and fire code regulations apply in a specific case, and how to tell whether appliances are installed correctly and safely.

Structurally sound houses don't wear out, but their component parts need to be replaced on a predictable basis. Plumbing, wiring, mechanical systems, roof surfaces, windows, and exterior skins (brick, stucco, stone, wood or other materials) have known life cycles. With regular maintenance and periodic replacement of necessary compo- nents, those postwar bungalows can stand up to repeated renovations and additions. But it takes a housing expert to determine the likely replacement requirements of any existing system, and to tell prospective owners what the costs will be.

While houses aren't boats, there are similarities in the transfer of these major properties. No mortgage company would provide the financing for a ship's mortgage without certain paperwork in place -- including a completed report from a qualified marine surveyor. This report must list the observed condition and current market value of the vessel, including a list of all equipment and its condition. In the housing market, financing institutions are becoming more dependent on two allied documents that provide a picture of the market value and the physical condition of the structure in detail. These are the home inspection report and the appraisal. As the market becomes more complex with buyers, sellers and bankers all dealing at arms' length, this new level of professional documentation is taking a permanent place in the transfer of property.

We are in an age of litigation, which creates anxiety on the part of buyer, seller, and real estate agent. Major new liabilities are being created or discovered through practice and by the various state legislatures. This creates professional opportunities for many different occupations -- particularly attorneys, and home inspectors.

And so the home inspection profession nationwide is in a growth stage. In just a few years, the percentage of home inspections done on the transfer of residences and other buildings will approach 90% or more. In addition to resales of existing dwellings, the inspection of new houses and construction underway provides a substantial market for the inspector. Because of tax initiatives and other limitations placed on the public purse strings, services are being privatized; builders are hiring their own inspections done in some areas to save time and money. That also spells an opportunity for people with the insight to learn the home inspection business today. This opportunity will lead thousands of alert, forward-looking individuals to a secure future dealing in one of the technical/professional fields to be born of this information age.

The home inspection industry today offers the chance of a lifetime. Men and women from different backgrounds can establish their own businesses to earn high incomes, with low front-end costs and very manageable overhead. While a knowledge of light construction, renovation, and skill in communicating technical information in a balanced and diplomatic way are essential, the field is open to virtually anyone.

Building and remodeling contractors are "naturals" for this field; also, retired military, educators, engineering and architectural technicians make potentially excellent home inspectors. Real estate agents, public sector building inspectors, draftsmen, journalists, and many others can find a better living inspecting houses than in their former occupations. There is strong demand nationwide for the skilled home inspector, and in many areas there is a real potential for a working inspector -- not necessarily as the owner of a business -- to earn $50,000, $60,000, or even more after developing the necessary skills. An annual income of $70,000 for a working inspector in the metropolitan Washington DC area is not unheard of!

An inspection service provides a careful technical analysis of the condition of existing buildings, using methodology that's relatively easy to learn. And unlike most other jobs requiring similar qualifications, home inspectors are enthusiastically appreciated for performing needed services that truly benefit their customers in major ways. The customer's perception of satisfaction is likely to be positive and high. A really good inspector will soon develop a wide circle of satisfied clients, and will enjoy top professional status in the community.

The services performed by a home inspector are generally completed in less than a half day, including preparation and delivery of a written report and collection of the fee -- which may range from $125 to $500 or more. While this fee obviously doesn't represent clear profit, it is easy to see that even a relatively slow inspector who produces only 10 or 12 complete inspections per week can generate from $1,250 to $3,000 weekly gross. Using streamlined reporting methods and a low-overhead office, the owner of a small home inspection business might expect profits of about 60-75% of the gross. And this is while producing a true community service, making new friends, and doing very satisfying work. If you establish an inspection business and operate it according to the HomeTech system, there won't even be a need for accounts receivable, because it will run as a cash business!

Most successful inspectors feel that the problem-solving aspects of the job are at least as satisfying as the pay. Inspectors not only identify hidden flaws, they show clients how to operate the house's various systems, how to anticipate needed repairs and maintenance, and how to prepare for the inevitable replacement cycle when things wear out. With relatively high pay, the satisfaction of having made significant discoveries to the customer's benefit, and the admiration of the client, home inspectors are geared for success. In the next few years, the market will be mammoth. By 1992, there will be a demand for better than 10 million inspections per year in this country. That translates to over 30,000 full time jobs as home inspectors -- four or five times the present number!

The home inspection field naturally dovetails with such occupations as home loan banking, home care, real estate appraisal, real estate sales, property management, relocation specialists, and small scale investment counseling. It accords professional status to the qualified inspector, regardless of the educational level attained in formal schooling. And while inspectors may be settled in their communities for life, with roots and a network of contacts similar to attorneys, this field also offers the individual a portable profession -- expertise as marketable in California as in Florida or Ohio. In addition to field inspection, it is possible for the well qualified inspector to specialize as an expert witness in courtroom situations, with a potential to earn $500 to $750 per day.

What Is a Home Inspection?

A home inspection is a controlled analysis of the present physical condition of a home, condominium, or other residential or commercial structure. It is a visual inspection which looks deeper than the surface to provide insights for customers that they would not otherwise easily obtain. This constitutes a technical service which carries with it responsibilities and certain legal liabilities, and puts the inspector in a position to have to pay for any oversights or wrong calls made. A home inspection, including customer conference and delivery of an on-site report, may take as little as two hours, but the impact of the report's contents may last for years -- along with the inspector's legal liability. Therefore the home inspection must be seen as a professional service rendered complete with a legally binding documentary report. There is no question that the inspector must be responsible, qualified, and very careful.

The home inspection has several components. First, of course, there is the actual process of data collecting with the inspector making careful notes based on a professional diagnosis of the house's condition. Then there is the "bedside manner" aspect, with the inspector as house doctor, telling the client in a reassuring way what is good and bad about the house. This requires a personable approach, and the inspector must genuinely enjoy interacting with clients. Next is the formal report -- either issued on-site and delivered along with a spoken commentary by the inspector, or delivered a day or two after the site visit has been completed. These reports may be handwritten on printed forms or they may be in letter form or another acceptable version.

The report, while only one aspect of the inspection, is the most lasting tangible evidence of the work that has been done. It is a powerful communication on an important subject, with implications that cut both ways -- to lessen the anxiety and reduce the unknown factors for the home buyer, and to define the limits of the inspector's responsibility, by using language that accurately describes conditions without unnecessarily opening the inspector to legal liability. Disclaimer statements contained in pre-inspection agreements are another useful part of the overall home inspection process.

Because of the developing nature of the home inspection business, there are various organizations offering franchises at fees ranging upwards from $14,000. While technical training and support are available through the franchise organizations, the value of the training may not be equal to the up-front cost -- and forms as well as other support materials may be updated irregularly if at all.

American Society of Home Inspectors

The American Society of Home Inspectors, Inc., (ASHI) a national professional society with headquarters in Washington DC, offers its members a variety of professional services. These include educational seminars, technical publications, a monthly newsletter, Code of Ethics and Standards of Practice as well as an annual national educational conference. Perhaps one of the most enticing features of ASHI membership is the availability of professional liability insurance. This is quite limited in scope, has a high deductible amount, and there are major doubts as to the long-term availability of coverage, but for the present it does provide some sense of security for individuals who are ASHI members.

ASHI membership is phased in, however, and requires from one to three years of candidate status before full membership with its range of benefits can be achieved. Standards are high, and in order to be considered for membership at least 250 to 750 fee-paid home inspections meeting ASHI Standards of Practice must be documented, among other requirements. Additional information is available by writing to ASHI, 3299 K Street NW, 7th floor, Washington, DC 20007.

Other national or regional home inspection associations have formed, and are forming. As of this writing, associations in existence include:

Home Inspection Institute of America, Inc.
20 North Plains Industrial Road, Suite 2
Wallingford, CT 06492

National Association of Home Inspectors, Inc.
5500 Wayzata Boulevard, Suite 1075
Golden Valley, MN 55416-1264

As with other professions, the home inspection field is a communications business. The same rules apply as with other professions:

√ There must be clear client/professional confidentiality.

√ The technical information must be of the highest possible quality.

√ The communication must follow a set procedure, with standards of practice and usage to sharply define the limits of responsibility for each party involved.

When and Where Is a Home Inspection Done?

Home inspections are performed on the site during daylight hours, year around. There are special circumstances when part of an inspection may have to be delayed -- such as checking the air conditioning function of a heat pump when temperatures are below 60 degrees -- in which case the inspection is completed with exceptions. The client pays, takes delivery of the report, and the inspector then returns when conditions permit to make necessary final checks.

How Are Home Inspections Done?

There are two basic styles of reporting used in the home inspection profession. A home inspection company must decide which approach it is going to take, train its inspectors accordingly, and develop inspection techniques that agree with this basic policy decision. The two approaches are as follows:

Technical Inspection. This approach is used by engineering-oriented individuals and companies. The model for technical reporting is the civil engineering profession, which is primarily confined to commercial and industrial buildings. The technical approach consists of an exhaustive inspection and analysis of every detail in a building. Some examples, perhaps extreme but valid nevertheless, include:

- Analysis of the heating system including calculation of the heat and cooling load of the building, such factors as insulation, window size, humidity and the like. This analysis might also include measuring the air flow at each register.

- Electrical inspection including calculation of the actual electrical load of all circuits in the house, checking voltage and amperage of each circuit, and taking off the cover plates at each outlet to check the connections.

- Plumbing inspection including measuring the sewer gases coming out of the vent pipe on the roof to determine if there is blockage in the line below. Also included might be measuring the water pressure coming in from the street, dismantling the hot water heater to check the condition of the tank, or measuring the time it takes for a bathtub to drain or for a commode tank to refill after flushing.

- Using a thermometer to check whether an oven heats up to the prescribed temperature in a given period of time, and to check the temperature of the refrigerator and freezer.

- If settlement or structural cracks are present, the analysis could include monitoring over a six-month period to determine whether the movement is ongoing.

Some of these tests seem extreme, but there is a school of thought in the home inspection business that believes this is the right approach, and that a home inspection should at least reach for these standards. This approach requires that an individual inspector be at least moderately well qualified in all specialized areas such as electrical, plumbing, heating, air conditioning, etc.

This type of inspector becomes almost "gadget-happy", and arrives at the inspection with a full set of tools and instruments to proceed with the inspection, which will typically require three to five hours.

General Inspection. This approach is the one that is most commonly adopted by inspectors throughout the country.

Most people who buy houses have little knowledge of home construction, and it is impossible to teach them enough in the course of an inspection so that they are able to evaluate for themselves all aspects of a property. This became clear when HomeTech was conducting a "How to Buy and Fix Up and Old House" seminar in Washington, DC. One full hour of the seminar was an exhaustive slide presentation of the integral parts of a house and how to inspect them.

At the end of one session, a student who appeared to have absorbed most of the information and had asked several intelligent questions about construction said, "Can I ask you one final question? Please tell me once again, what is a double-hung window?" This was proof positive how little a typical homeowner can absorb in the course of a 4-hour seminar, much less a 1 to 2 hour home inspection where they have many other things on their mind.

Most home purchases are made more on emotion than on logic. People are not buying a house with all its technical characteristics and components, but are in fact buying a way of life or a lifestyle that a house will provide them. Investors are making a capital investment related to dollars, not a technical decision.

An enormous number of people say such things as, "I really bought the house because of the beautiful tree in the back yard outside the kitchen window," or "It is my understanding that Mr. Smith who was the Secretary of the Treasury lived in this house. Isn't it wonderful to live in a part of history?" or "Isn't the effect of the family room stupendous?", talking about the furnishings rather than the room itself. Purchasers often fail to recognize basic elements of a house which would have an enormous effect on their purchase and liveability.

Going through a house with the home inspector who points out sizes of rooms, floor plans, as well as maintenance features and expandability potential, allows them to put things in proper perspective and make a more knowledgeable decision on whether to go ahead with the purchase.

A home inspector must talk about both the good and bad features of a house, put these features in perspective, and give purchasers the facts they need to make decisions on whether to go ahead. It is not necessary to explain the operation of a heat pump or spend 15 minutes in a discourse on gas versus electric cooking. What most purchasers want is that the inspector will check all the major integral parts of the house, point out the salient facts about the property, and give them his best evaluation of these items.

Most home purchasers, particularly first time purchasers, do not know that every house has problems and that it is impossible to find the perfect house. A home inspector is going to be able to find something wrong with almost every property that is inspected. And if a house is close to perfection, it is likely to be so expensive compared to the market that it is not cost-effective for purchasers to pay for that "perfection."

While the home inspector points out defects of the house, it is important to explain that every house has some problems, and that purchasers must look at the overall good and bad aspects of the house to make a decision.

Some years ago a real estate agent was lamenting the fact that an inspection company had killed three deals for her in the previous month. Her concern was not what you might expect, however. She put it this way, "Joe Miller of the XYZ Home Inspection company killed three deals for me last month. It made no difference to me as I turned around and sold all three clients other houses, so all it did was take me a little longer to make the commission. What was disconcerting however, was the fact that two of the three clients told me that, having looked at many more houses between the first and second contract, they wished they had not kicked out the contract on the first house. As they said, they did not realize that every house has some defects and you have to look at the overall picture." This kind of story has been repeated hundreds of times by real estate agents.

As another successful realtor said, "I insist an inspection be done on every house I sell, but I can't stand dealing with alarmists who magnify every problem and scare the buyer away from going ahead with the contract. That type of home inspector can kill any deal on any house by magnifying defects and not recognizing the house for what it is and putting all factors in perspective."

What purchasers want is a construction professional who will give them the facts they need to make a decision about buying a house: the good and bad, the strengths and weaknesses of a property, and how to evaluate them in the overall housing market. This approach is superior to the technical approach. A technical approach must be evaluated by a technician, and most home buyers are not qualified to do that.

Several years ago HomeTech participated in a seminar given by the Board of Realtors at which 12 different inspection companies served on a panel. Each company was given a chance to speak, and most companies gave their qualifications as well as their fee structure. One inspector, by way of demonstrating his thoroughness, explained that as part of the inspection he brought along an instrument to check the gas coming out of back vents on the roof. In this manner he was able to check whether there was stoppage in the piping. He felt that this was a benefit.

When the time came to explain about HomeTech, we concentrated on how the home inspection benefits real estate professionals, trusting that they knew enough about HomeTech to select us from the field. About the previous inspector we said, "If you want your clients to get that kind of inspection, call him; we do not do that," and went on to explain that we were generalists who provided an overview of the house spelling out strengths and weaknesses and putting a property in perspective. As a result of that, we got many inspections.

It is important to differentiate between minor items and major items. One of the advantages of requesting that purchasers be present at the inspection is that it is easier to provide perspective face-to-face than in writing or on a checklist. If the purchasers are not present at the inspection, it is wise to call them on the phone as well as doing your best to be clear in the written report, to provide the proper perspective about major and minor defects, and the strengths and weaknesses of a property.

While different individuals will develop variations to fit their own personalities, the "how" of the inspection process rests squarely on four things:

√ The inspector must take charge of the procedure, and not just follow the customer's agenda.

√ The inspector must work efficiently, which means using a familiar technique and method.

√ The inspector should take two passes through, so items that may have been missed or questions raised on the first pass can be caught or settled on the second.

√ A good inspector will create a context for the information, to give the customer an understanding of the importance of what is discovered.

Why Home Inspections?

Buying or selling a home brings up hundreds of questions that don't occur in the daily life of most people. A few of the questions center on the structural condition of the building, requiring the evaluation of a building expert or home inspector. Other troubling concerns during the time when a home inspection might be ordered can range from the quality of the neighborhood schools, to security, and major financial/employment questions. The home inspector is a specialist employed during a time of stress, to help identify strengths and weaknesses and to point towards a solution.

Because the home buyers or sellers are not construction specialists, they may not realize that a wet basement may be remedied with a few hours' hired labor or minor regrading. To the impressionable person, minor flaws can appear major and can scare away buyers from an excellent home which needs a few cosmetic repairs. To present a clear word picture of the house in realistic context, the home inspector must deliver an expertly crafted report. Presenting reassurance in a time of doubt and trouble earns the inspector a special status in the eyes of the customer. If the report is accurate, that high regard will last indefinitely. The home inspector had best be certain that the report is correct.

The home inspection report can pinpoint necessary repairs to save the buyer thousands of dollars, especially if local laws are consumer-oriented. Electrical and mechanical in-working-order clauses are routine contract elements in many jurisdictions. When an inspector finds a defective furnace, an unsafe wiring system, or a plumbing system in need of major repairs, the result can be a reduction in price to the home buyer -- and a savings of whatever the repairs might cost.

However, if the inspector overlooks an item that is covered in the "working order" clause of the sales contract, the inspector or company providing the inspection service can be held liable. Sloppy inspections sometimes lead to major errors and omissions. That can translate to home improvements paid for in part or entirely from the inspector's personal funds.

Home inspections may be ordered for a variety of reasons, but most can be reduced to the following:

√ Structural and mechanical problems are identified, and solutions are specified.

√ Buyers' protection -- money saved in hidden or latent defects which can be held the sellers' responsibility to make right.

√ Discovery of existing conditions that can provide a "way out" of a sales contract.

Mission of a Home Inspection Business

There is a slight difference between the mission of the home inspection business and the reasons listed above for a buyer to call for an inspection. As an information service organization, the product offered is technical information based on wide knowledge and general familiarity with housing -- construction methods, the nature of components and materials life cycles, current conditions in the labor market, and the cost of materials and equipment.

The mission of the business is threefold:

√ To provide prospective purchasers with the facts and information they need to make an enlightened decision about purchasing a home or building.

√ To provide real estate professionals with a service that will enable them to properly assist home purchasers in making a knowledgeable decision.

√ To be an information source for home buyers, home owners, and real estate professionals regarding the components and systems of the home -- its maintenance, remodeling, and renovation.

From a business viewpoint, the real estate salesperson is the source of most calls for home inspections. It is important to establish a rapport with the real estate industry and to assure local agents that the service has strong benefits for them. In some instances, agents may feel that there is an adversarial relationship with the inspector, especially once an offer has been signed and the contract written. This feeling is not always wrong.

Some inspectors are seen as nitpicking deal-killers because that's the attitude they project. And while it might give an individual a sense of power to cancel a $100,000 contract with incisive comment and pointed findings, the inspector with such an attitude will not be long in business. The word will circulate in the community, and there won't be many repeat calls from agents whose commissions were washed away.

The HomeTech method maintains that honest, professional real estate agents want buyers to be fully informed and able to make the best possible decision. If a house has flaws -- and what house doesn't? -- it is the inspector's responsibility to make the findings and then present the information in a positive light wherever possible. As dedicated as the inspector must be to accuracy, the integrity of the contract must be preserved. Fairness is the watchword. It can be the key to success or failure of an individual's career in home inspection.

In Texas, the state licenses home inspectors and controls the quality and methodology used so that a theoretically uniform system of documentation is assured. Sellers pay for repairs up to $750, and an arbitration system allows buyer and seller to bring a property into reasonable condition as a condition of sale. In California, there is a "latent defects" law that makes sellers or their agents liable for most major defects that may be discovered in a property for several years after the sale. This has caused panic in the real estate industry and an overwhelming demand for home inspections, the inspectors becoming at least partially liable for any defects that are not disclosed at time of sale. All the more reason for inspections that are thorough, accurate, balanced, understandable to the buyer, and timely.

Legislated certification of the home inspection industry is coming. It could help promote home inspections by educating the public about the need for that service. One danger of legislation is that it might establish minimum standards which then, in turn, would flood the market with minimally qualified inspectors working from a checklist specified by the state, which will depress the fee value of the individual inspection. This is a sort of standardization which could demean the profession generally.

It is critical, therefore, that inspectors set high standards now and establish a level of performance and general practice acceptable in the insurance and financial marketplace. Only by offering and delivering a predictable quality product will the industry be in a position to avoid the "lowest common denominator" effect of public sector certification. Professional home inspectors and organizations such as ASHI will have a chance to participate in development of licensing standards.

Cast of Characters

Viewing the home inspection process as a dramatic production may set the stage for gaining an understanding of the various characters who will populate your business day, and how each may act and react. As an inspector, you are a communicator and your main activity can described as education. Here are the characters you're likely to meet, and how you should address the needs and expectations of each one.

Purchaser

In nearly all the jobs you'll inspect, the purchaser is your client. The relationship you will maintain is confidential: you're working exclusively for this client and not the seller, the sales agent, or anyone else. The report that you generate is the client's property and should never be released to anyone else without the expressed permission of the original client.

If at all possible, **the purchaser should be present at the inspection**. There are several reasons for this. First, the purchaser will see you in action and realize the value of the report they are left holding after you've moved on. Second, you'll have a better opportunity to educate the clients about the care and operation of the house if they walk with you on your second pass through.

You'll be able to develop a mutual understanding that requires a two-way communications process, which in turn helps you establish the context, or weight to be assigned to the data discovered. If your inspection must be made in the client's absence, your liability increases because there is no opportunity to develop rapport, and everything depends on the interpretation of the written report.

Understand that every good home inspector is a bit of a ham, putting on a "show" for the clients. Getting down on your hands and knees to look up a chimney flue, going into difficult crawl spaces, and climbing up into attics is not only necessary for a thorough inspection, it is very reassuring to your clients.

You may encounter several different types of people with various needs.

First is **the handyman client,** often a blue-collar type who wants a thorough structural inspection, to determine whether the house is going to require major repairs or cosmetic improvements. This client usually wants to be reassured that the house is worth his time and effort, and he can take care of the repair and maintenance. This is the client who will say, "I can fix anything. Just tell me about the major systems and whether the house is going to fall down."

Your professional analysis will be carefully considered, but the blue-collar client is less inclined to deal with professional reports and analysis than the typical investor would be. More comfortable with immediate and practical solutions than with theory and planning, this client will be helped most by your careful analysis and clear how-to explanations.

The investor will want to know the current status of the property, but also will be very interested in the cost of ongoing maintenance and repair, the "X plus Y equation" (see page 35 for more about this). This client will ask questions primarily from a financial viewpoint and may want your opinion on the value of the property. Beware of answering that question in specific terms: it could create an additional liability for you. Instead, reply in general terms such as, "it appears to be in the ball park compared to other properties in this area."

Your commitment is to provide assistance in the physical/structural area, and you should be prepared to offer a comfortably wide range of costs for replacement/repair of the major systems in the house. You must make your statements general enough so that you're not pinned down to a specific figure. Investors want to know approximately what their costs will be for the time that they expect to own the property.

You may also encounter a **homeowner resident** as a client, who calls you in for a diagnosis as a "house doctor". This might be an elderly lady whose husband has died, and is moving into a smaller house, on a tight budget. With this purchaser, even a minor plumbing defect or washer that needs to be changed will require calling a plumber with a minimum charge of $45-$50 plus the cost of the repair. What this purchaser wants most is a thorough inspection of the house not only to ensure there is nothing major, but to ensure the house is in tip-top condition, and there is no deferred maintenance that is going to come up in the future that is going to throw her house budget out of whack. In this instance the inspection would be much more critical of minor items.

Keep moving. Tell the clients you've got to make a methodical pass through the property and that they will be welcome to ask questions or accompany you on your second pass. Then you can lead the discussion by pointing out certain things in a positive way, and giving answers to a variety of possible questions.

Renovator/owners may be another type you'll encounter. These folks are similar to the homeowner resident, with specific questions. Will the structure take a whirlpool bath in a master suite in the present attic space? Is the wall between the kitchen and dining room a bearing wall? Can it be eliminated entirely? What sort of support is needed? How to control the mildew in the basement, and what about those sagging rafters and the roof with its layers of shingles? How should we repair the place to stop deterioration in anticipation of a two-phase remodeling plan?

Remember that your services are not those of a professional architect, engineer, electrician, plumber, grading contractor, or heat/air conditioning contractor. You can provide analytical insights. You can make an educated guess as to the age and condition of various systems, probably much closer to the facts than most of the other experts named. You can "see through" the house based on your experience with other, similar places. Your technical opinion will be expert, but be sure you present it as that in your report. Avoid specific dollar figures, but offer ranges instead.

There are home inspectors who will not even give a range of costs because they are concerned about potential liability. HomeTech contends that cost ranges are valuable to a client who needs information to make a decision on a property. Knowing the approximate cost of finishing a basement into a family room, or enclosing a porch for a mother-in-law suite, can be critical in a decision to purchase.

Real Estate Agent

This character is in control of the home inspection industry today. Home inspectors pay courtesy calls on major real estate offices to develop a cordial relationship and, frankly, to pick up orders. While the client pays your bill, it's often the agent who chooses which home inspector to use. This key person is part of any successful home inspection business strategy.

Probably 25% of home inspectors believe the real estate agent is an adversary. HomeTech recognizes the agent as an influential professional worthy of respect, and assumes the agent's interest is best served when the buyer has the information needed to make the best possible decision. While agents legally represent the seller, they make arrangements for an inspection as a post-sale service. The best attitude for the home inspector is to give, and expect, fair and impartial treatment.

Because of potential liability, real estate agents frequently do not recommend just one home inspector, but instead give the purchaser a list of three or more inspectors. Even so, the smart real estate professional has considerable influence on selection of a home inspector.

The agent should be present along with the purchaser at the inspection. The agent's job will be to handle the seller, giving the purchaser freedom to investigate and ask questions of the impartial expert, the inspector. The agent will usually have the purchaser's confidence and can help give perspective on reopening the contract for negotiation on minor points. Working as a professional team, the inspector and the agent can help the buyer gain self confidence in the face of this major commitment.

"CARE AND FEEDING" OF THE AGENT

A clear business relationship should be established early on. Payment of a commission or any other direct consideration such as a gift of any kind is not appropriate or professional. The courtesy of an occasional free inspection when an agent buys property personally is considered reasonable and fair.

The home inspector is engaged in an educational process, so it is important to offer some special services. These may include slide presentations at sales meetings, speeches to the luncheon meeting of the local Board of Realtors or other groups such as civic clubs, the Chamber of Commerce, etc. Your business source is the real estate industry, and your referrals will come through agents and real estate attorneys.

HOW THE HOME INSPECTION HELPS MAKE THE SALE

You can outline several situations where a home inspection is a positive selling tool.

The "Nervous Nellie" buyer. After having seen a number of properties without reaching a decision, this buyer can often be persuaded to sign a contract with a contingency based on a home inspection. The deal usually goes through, unless something major is discovered in the inspection.

Assessing an obvious known defect. If a house has an obvious defect, such as a basement water problem or a bad roof, prospective purchasers may shy away or make a very low offer that more than covers any possible costs. In this case, your expert analysis and estimated cost will put perspective on the situation.

Estimating substantial remodeling or renovation. This is a job some homeowners might expect a contractor to do, but an inspector can analyze the situation and provide a guideline for the client. The information will help the real estate agent to create a detailed picture of the finished house -- and convert a prospect into a client.

When purchasers are at or near their financial limit. In many cases the cost of monthly payments places a family so close to the line that any unexpected major repairs could force them to the brink of foreclosure. An inspection can identify likely capital improvements required in the near future, and this unfortunate situation can be avoided.

Third-party kibitzer. A deal is about to be signed when a relative from out of town or a friend in some field related to housing walks through the house with the prospective buyer. The so-called "expert" may not be entirely versed in the conditions: maybe the price if this friend is from a different area. An impartial home inspector and report can furnish a credibility that the real estate agent can't -- and it could save the sale. When things go as they should and inspector can talk to client one-on-one, fears can usually be eased.

These arguments can help convince real estate agents of the value of a home inspection. Only by establishing a mutual trust, and a reputation for dependability and fairness, can the business relationship be fully developed.

Seller

If all sellers were 100% honest and informed, there would be no need for a home inspector. That's a slight overstatement, but the fact is that sellers can pose the greatest problem for the inspector. You can't very well deny the seller directly when he asks, "D'ya mind if I go along?" But you can avoid incidents by being innocuous, going through your routine and keeping quiet about what you see. At the end of the inspection, sit down privately with the purchaser and give your report.

The worst situation develops when an inspector says a water heater is 7 to 10 years old and the seller chimes in that it's no such thing, that water heater is only 6-1/2 years old this month. This disruption can damage your composure, detract from your reputation, and harm the prospects for a sale. If a seller insists on being at the house during the inspection, enlist the real estate agent's help to keep the seller occupied until you make your inspection and give your report verbally to the purchasers in private. Only then are you free to do the job the purchasers have hired you to do: discover the existing conditions and put matters in perspective.

Real Estate Lawyer

The second most likely source for inspection assignments is real estate attorneys. Usually on the buyer's behalf, the attorney may send you assignments and indirect referrals through the fraternity of the bar. Your best strategy is to provide high quality work, be dependable, and write a report that anyone can understand.

In some instances the attorney will prefer to handle the assignment through a secretary, and pay the bill for services rendered at settlement. If you want the business, accept the terms. There's nothing unusual about waiting a little while for a professional fee, and since the attorney's office will cut all checks for distribution at one time, you won't have to carry a large accounts receivable on your books for that reason. While home inspection is essentially a cash business, there are exceptions. Just be sure, if you're delivering a written report to the attorney's office, that your statement for services rendered is prominently attached. Remember, too, that it is always a good idea to request that the purchaser be present at the inspection.

Real Estate Appraiser

The fields are allied and similar, but they are separate. You can send appraisers business and they can return the courtesy, but you should be very careful to avoid stepping into the appraiser's arena and offering an opinion on the dollar value of a property. A whole different set of circumstances, liabilities, and practices are involved.

You can identify the condition of the house, and the condition and repair or replacement cost of integral parts that may require cash expenditures in the next one to five years. Thus, while an appraiser will be able to assign a present market value to the property, your numbers are likely to be smaller in scope but no less important in creating a realistic financial picture. In some areas of the country, where double-digit annual property appreciation is still commonplace, the repairs you list will probably be more than covered in one year by the increase in property value. But in many areas rising property value is no longer a certainty, so the cost of repairs is more of a concern to purchasers.

Summary

Home inspection is a developing occupation which presents a great opportunity today. A large part of the American housing inventory is over 20 years old, and technical changes that have taken place since the end of World War II have combined with social trends to establish this new profession. The opportunity is there for individuals with relatively small capital investment to achieve better earnings than they have ever known. The market for home inspections is growing rapidly, likely to increase to more than 10 million inspections per year and 30,000 full time inspectors in the next decade. Inspectors earn professional status and pay.

A home inspection looks below the surface to identify physical/structural condition of a house or condominium, but more is involved than simple data collection and reporting. The manner in which the data is reported must be defined and controlled, for the protection of the inspector and service to the client, who is distracted with other concerns. A set methodology is the best approach to the inspection process for several reasons. These include keeping the inspector in control of the process, time efficiency, and creation of a consistent style of communicating the results of an inspection.

Three key words are liability, accuracy, and clarity. Liability is the result of insufficient accuracy in observing conditions, and lack of clarity in communicating those conditions to the client. Inspectors can be made to pay for their oversights.

The home inspection business is an information source for home buyers and the real estate industry. Communicating the condition and future repair needs of a home will increase the purchaser's awareness and can help nervous prospects to sign a contract offer. The inspector can help the real estate agent as well as the purchaser.

Conversely, the real estate agent controls the home inspector's calls for jobs. Inspectors must develop a working relationship with agents and real estate companies. This involves sales presentations and slide shows, but the lines of professional courtesy don't include cash payments between inspector and agent as a means of securing work.

A number of other characters populate the inspector's business world. These include the purchaser, for whom the professional inspector usually works. There are several types of purchasers. Also included are the agent, the seller, real estate attorney, property appraiser, and others. Knowing the types and their potential needs can help the inspector plan a service strategy, which is necessary in order to be most effective.

2. QUALIFICATIONS AND STANDARDS

Home Inspectors Today and Tomorrow

Qualified home inspectors come from all walks of life and all age groups. Today's average home inspector is a middle-aged white male with a background in construction, remodeling, engineering or architecture. Often he's in a second profession, retired from an occupation of higher stress -- perhaps struggle to make a small business work.

Most commonly, a home inspector is a "burned out" remodeler or builder, with good construction knowledge, a well-developed work ethic, and ability to deal with people. He now counsels home purchasers from the perspective of a "Dutch Uncle" with mature judgement, enjoying better cash flow and lower stress levels than he's ever known before. Little wonder that he loves this work: it affords community respect, moderate financial independence, and the added reward of providing a much needed and significant service to people at a crucial time in their lives.

Being a professional friend to people who are in the throes of relocation, made insecure by imminent changes in routine, can be as satisfying as any role in society. The inspector is a detective, technical advisor, and a friend.

He's a house doctor with a good bedside manner.

He's also a businessman in control of his own situation. With two or three inspections daily, he can earn a very respectable net income of more than $50,000 annually. Various personality types with different financial expectations can find a comfortable place in this field. Today's average inspector can choose a semi-retirement lifestyle and supplement his income with $30,000 additional gross by doing just three or four inspections a week. He can work until he no longer feels comfortable with the mild physical strain of climbing a ladder or crawling under a house. Life will probably be kind in his golden years . . .

But make no mistake -- the opportunity is drawing attention and attracting new blood. Younger inspectors are seeking careers, and over the next decade they will dominate the ranks. Their higher energy and greater ambition will bring in more aggressive business techniques, stronger emphasis on methodology, and more competition. The result will be better quality inspection services and generally higher standards industry-wide. This will require executive abilities and disciplines, and your future will be shaped by mastering these skills.

Attractions of the Home Inspection Business

Home inspections can be a lot of fun. If you like people and you like houses, it is an ideal line of work. Every home inspection usually involves at least two purchasers, a listing and selling agent, and in some cases a seller. So you are meeting a great many different people and helping them at a very important time in their lives. Also, just as each person is different, every house is unique, and you will constantly run into different conditions which require attention and allow you to engage in problem-solving.

The inspection business is what we call flexible/inflexible, in that you do not work on a 9 to 5 basis. At the peak of the season, if business is good and the market strong, you could work sunup to sundown if you wish. At other times of the year and around the holiday seasons, the demand falls off and there is more free time. While the hours can be relatively long, and the work is considered hard by some, when you are finished with an inspection and the report is completed on-site or dictated, you are through and there is no homework to be done. At most, you might have a follow-up telephone call or a recheck if problems arise.

Anyone who has been in the remodeling business will appreciate the relief from relentless pressure. You can start and finish a job in 1 or 2 hours, then you are done. If you have a cancellation or a slow schedule, and you complete your work at 2 PM, then you can go home and relax.

You can start in the business on your own, or join a company on a part-time basis and continue your other employment. As your reputation develops and business increases, you can change to a full-time venture. On the other hand, if your health is good and you wish to slow down toward retirement, you can do inspections as long as you are physically agile.

Whether you work for yourself or for another company, the pay is relatively lucrative.

The home inspection business is not very capital intensive. Although a great deal of money can be spent in starting out, most commonly the greatest investment is in time, not money. There is no investment for equipment or showroom, and you can work out of your home. There is no large staff to hire and train. Thus, you can start up in business with very few up-front dollars.

Job costs are predictable and easy to control. The only direct job costs are the inspector's fees, and if you are running the business and doing the inspections yourself, there is little cost involved. You can predict what your income will be as the jobs come in. This is in direct contrast with the remodeling business, where estimating job costs and controlling job costs is a prime consideration.

The only way you can get hurt in this business is that you might spend 2-1/2 hours on an inspection rather than 1-1/2. If you hire other inspectors, the general practice is to give them a percentage of the fee, so whether you are paying them as employees or subcontractors you are working on a fixed-price basis.

The business has excellent cash flow, because nearly all the time you are paid immediately upon completion of the inspection. Simply tell people that payment must be made at the time of inspection by check, cash or credit card. You take a 3%-5% discount on credit card sales, but your cash flow is regular and there are no accounts receivable to collect.

Production is easy to control. If you are running the business and doing the inspections yourself, you only promise what you can produce. If you are working with other inspectors, they are usually professionals who are conscientious, capable, and will perform as promised.

This a real contrast to the remodeling business, where one of the biggest problems is that your word is only as good as the performance of your crews and subcontractors. You are constantly forced to make commitments and promises based on other people's performance, and it is not always possible to guarantee their production. In the home inspection business this is not a problem.

The home inspection business need not be overhead intensive. If you are doing all the inspections yourself, you can work out of your home, use an on-site written report, and get by with an answering service. Thus, your budget is low and in slow times you are not eaten up by overhead. As your business grows, and you add inspectors, your overhead can go up as high as 30%-40%.

The home inspection business allows an individual to branch out and develop other interests, vocational or not. Because of the ability to control time and have flexible hours, a home inspection business can be worked in with teaching, coaching or avocations such as playing tennis, golf, etc.

There is a tremendous demand for people who are in the information business related to construction, real estate and renovation. Thus, a home inspector can develop numerous sidelines such as writing articles, giving seminars, teaching classes, providing expert testimony, consulting or even buying and fixing up investment property.

Finally, the home inspection business provides a great deal of psychic income. Home inspectors are providing a badly needed service to home buyers and are helping people make what is probably the largest financial investment of their entire lives. For many prospective buyers this is a time of crisis, and dealing with a competent, self-assured inspector can be a source of great comfort. As a result, inspectors receive compliments and accolades on a daily basis, with adversarial relationships kept to a minimum.

Qualifying as a Professional

To qualify as a home inspector you must have a detailed understanding of residential construction, but you don't have to be an engineer or architect. You must have what we call a "frame of reference" about how a house is built. With this knowledge as a base, you then learn the specific information to be a qualified home inspector.

You must be able to communicate your understanding in a balanced report that non-technical people will easily understand. This means you must draw on verbal skills and people skills as strong as your technical knowledge. While there are no organized apprenticeship programs in the field, it is generally necessary to spend several months to a year doing regular inspections according to a consistent methodology before you're eligible to be considered a seasoned home inspector.

Once you've mastered the HomeTech method and have the required skills, it's a matter of being exposed repeatedly to the realities of different kinds of characters and meeting their needs -- thinking on your feet -- while carefully crafting your report strategy.

Only field exposure will give you the experience necessary to perform as a professional home inspector.

ASHI Standards

The American Society of Home Inspectors, Inc. (ASHI) has stringent requirements for membership in the first national professional society of its kind. In some instances these requirements make it difficult for well-qualified inspectors to gain active membership, a situation which has held down the membership rolls considerably. To become an ASHI member one must:

√ Perform at least 250 fee paid home inspections within a 3-year period that meet or exceed ASHI standards

√ Complete a minimum of one year as a Candidate in good standing, performing home inspections in accordance with ASHI standards and the ASHI code of ethics

√ Pass a series of three written examinations

√ Continue to comply with ASHI Code of Ethics, Standards of Practice

√ Complete ASHI-approved continuing education credits, 40 every two years

Members are given the privilege to use the ASHI logo in advertising, benefit from public relations programs, vote and hold office in the organization, receive a monthly newsletter, and have access to technical papers, conference proceedings, consumer brochures, surveys, etc. Access to ASHI-sponsored professional liability insurance is one of the advertised benefits of membership, but the cost and coverage are of questionable benefit. Such policies generally have high deductibles and many exclusions.

Other home inspection associations in existence at this writing appear on page 6.

Educational Background and Training

It's good, though not essential, for home inspectors to have a college career. A technical degree is helpful because it adds to the professional qualifications and may be useful in your practice of home inspections. But a liberal arts background, in which verbal and thinking abilities are emphasized, is equally valuable. While a bachelor's degree in sociology inscribed on a diploma hanging on your office wall may not say a great deal about your construction knowledge, it will help you in the actual conduct of your home inspection career.

Consider these accepted conventional ways to obtain the construction knowledge you'll need to become an entry-level home inspector. You could have:

✓ Five to 10 years experience in home building, remodeling, or renovation;

✓ Earned a college or trade school degree in architecture, engineering, industrial studies, or construction technology accompanied by on-the-job experience;

✓ Learned by working in the construction field or as a draftsperson or other information specialist relating to home building, plus on-the-job experience.

Necessary Personal Skills

An inspector must know more than construction theory because so much of what must be uncovered and described relates to how buildings weather and how equipment ages. Knowing current standards of practice in foundation design is useful, but the inspector who comes upon a 200-year-old farmhouse with some evidence of foundation settlement has to be able to make sophisticated, educated guesses in several directions at once. Then he must combine the assumptions to report to clients who know -- and care -- hardly anything about Colonial foundation practices. The report won't contain all the inspector knows on the subject, but it must present a range of costs and therefore the inspector must know how to solve almost any problem.

PROFESSIONAL IMAGE

It is important to have either education or experience that makes for an articulate stage presence. You must be able to talk intelligently about the economy and real estate markets as well as about construction. As a technical advisor you are routinely consulted regarding the advisability of purchasing a property worth from $75,000 to $250,000. It is essential that you be able to function as a peer of the prospective purchaser. As a home inspector you are not a tradesperson; you are a professional who is willing to inspect houses.

Identifying the problem is basic. Suggesting a solution that is realistic, and even offering alternative solutions, is the arena of the seasoned professional. Skill lies in doing this without being overly specific; pointing directions and giving ranges, rather than providing specific how-to information.

In addition to detecting structural problems, you must know the basics of electrical, heating, plumbing, and insulation code requirements. On-the-job experience is the most dependable teacher. Seeing the actual conditions makes a more lasting impression than reading text, or memorizing data.

EMPATHY

You must be able to empathize with people. You'll be meeting with people for an hour or so several times in the work day. You must relate to the client's needs, discuss concerns and allay fears, and explain the house analysis in clear and simple language. Lacking these skills, an inspector can know the date and method of manufacture of every timber, wire, pipe, nail and shingle in the house and still not succeed. This is a communications art. Data collection is one of the main components, but people skills are equally vital.

Skill as a public speaker or slide show presenter is very desirable in a home inspector. This helps develop the market, gives the inspector a chance to demonstrate personality and establish an image of knowledgeable humility. It also helps project the image of the company as being fully involved in the business community.

SELF CONFIDENCE

Since inspectors are paid on commission, legally making them subcontractors and not employees, they are essentially working for themselves. It is important that the individual have self confidence and an entrepreneurial attitude. People who have relied on the security of a regular salary might find it difficult to adjust to the fluctuating work pattern in the inspection business. This needs to be explored in depth with any prospective home inspector, whether you are considering getting into the field or are looking for inspectors to hire.

THIRST FOR KNOWLEDGE

A good home inspector is a perpetual student eager to attend technical seminars, read magazines about housing and construction, study the real estate section of newspapers, and stay aware of what is going on in the business. Housing and real estate is a dynamic field. For example, between December 1985 and January 1986, asbestos went from being no call in a home inspection to perhaps the most important call, with tremendous liability implications.

Home inspection is not like construction where once you learn to hang a door, you know that forever. The business is constantly changing, and only those people with an appetite for learning can keep up.

Standard Paperwork

For your own convenience and protection, it is recommended that you use a consistent format in operating your business, conferring with clients, and making your inspections. HomeTech has developed a methodology to serve this need, and using it will increase your confidence (it's a proven professional method of operation), protect your assets (avoid unnecessary liabilities), and give you the ability to reach your own cash flow goals.

You can use printed forms from HomeTech to document your entire operation.

Pre-Inspection Agreement

One of the most important forms to consider is the Pre-Inspection Agreement. This describes what will and won't be covered by your inspection. Terms used in this agreement may vary over time, because legalities and liabilities are changing continually, but the concept is to explain your service clearly to clients, so they will know just what to expect from the home inspection.

Inform the client that your inspection is made to industry standards, you do not guarantee the condition of items, but you are rendering a professional opinion based on a visual inspection made on the specified date. While most equipment is durable and long-lasting, any specific item could fail at any time. Your inspection describes conditions observed at a point in time, and projects probabilities. Speak in terms of probable time periods, giving ranges for replacement costs and budgeting required for contingencies.

You make your sale by creating confidence in your professional ability and experience, while at the same time shaping your customer's expectations. This creates a context in which you can comfortably work. It also tends to limit your liability and lessen the chances that the clients will turn on you for repairs if something malfunctions.

While the pre-inspection agreement may not prevent every possible complaint against you, it will go a long way towards protecting your pocketbook. With that document signed, your job is clarified and all you need to be concerned with is delivering the best home inspection of which you are capable. Don't even think about performing home inspections without a pre-inspection agreement.

Inspection Report

Having made the physical inspection, collected all notes, and conferred with the client, you'll take one of three parallel avenues to finish the inspection process. The measure of your success -- and financial reward -- will depend on which track you choose and how well you pursue it.

No matter which reporting method you use, the first rule is to **complete your report before you go on to the next inspection.** You cannot do several inspections in a day and then dictate or write out the reports. If you try, you will find that your accuracy suffers greatly and you spend a lot of time trying to recreate what is no longer fresh in your mind. You also lose one of the benefits of being a home inspector: when your day's work is done, you're done.

CHECKLIST REPORT

This report is delivered at time of inspection. The HomeTech Building Analysis Report (BAR) is designed to help you proceed methodically through the inspection without wasting time or missing important data. This is an 18-page NCR (carbonless copy) document which you fill out as you make your rounds, keeping a copy for your own records and delivering the original to your client.

The form provides standardized language to describe the inspection process and limit your liability. It gives the inspector a "script" to follow in making a verbal report to clients, and it provides a short "owners' manual" telling how to deal with routine home maintenance. In addition, there is a handy reference showing cost ranges for repair and replacement items. This BAR is helpful to inspector and client as a medium of information delivery. It gives the client printed reminders on how to solve common problems.

Your job is made simple and clear, but you must present the emphasis and shading that makes a reassuring report. You don't just fill in a checklist and hand it over. Your professional interpretation of the printed matter is contained in your handwritten and spoken delivery.

This is the fastest way to make a complete inspection, deliver the written report, collect your check and move on to the next job. More important to the clients, this expeditious reporting provides the information necessary to decide whether to release the inspection contingency on the property. What the report may lack in professional appearance is more than made up by its timeliness, which makes it almost universally preferred by real estate agents as well.

DICTATED LETTER REPORT

If you use a narrative letter report, you can save time on the site by giving the client a short summary report at the time of inspection. Keep a copy for your files and to shape your dictated letter following the client conference. Time on the site will be somewhat less than using the BAR, because you won't have to fill it out in order to hold your conference. However, additional time in the office is required.

HomeTech's short on-site report provides the client with an immediate record of major items discovered in the inspection, and organizes your notes into consistent format for dictation. This report is used as a preliminary record only, and is followed by a full written report. The dictation for a narrative report is most effectively done immediately after you leave the site, while in the car on the way to your next appointment. With the summary report, reasonable dictation skills, and dedication to getting the job done right, you can bring in a cassette tape at the end of the day with three or more valuable narrative reports.

If it takes an hour of secretarial time to prepare a dictated report and mail it, compared to 15 minutes of your own time filling out the BAR, there is a basis for cost comparison. If you're out on the road all day and return to the office in the evening with a tape of dictated reports, the earliest possible time you can offer delivery is 24 hours after the inspection is made. And that assumes the client wants to come to your office to make the pick-up -- or can receive the report by fax machine if you have one. If you rely on the U.S. Postal Service for delivery, you can figure from one day to several before delivery is completed.

Cost for the narrative report is one consideration. The quality of service -- how long it takes the client to have a document that can be used in negotiations -- is a second factor. Many inspectors prefer to dictate a narrative report simply because they feel the communication is more complete. In a letter you can say exactly what you mean to say.

Letter reports have several drawbacks for the inexperienced home inspector. Proper disclaimers, correct wording on various issues, and a consistent form should be used in all letter reports. Until you have made so many reports that the words come naturally to you -- that is, until you've conditioned your mind to "read out" the phrases semi-automatically, the letter report will be clumsy and may not protect you from liability.

Advantages are: precision and personal communication. Disadvantages include: cost, time of delivery, possible errors or omissions resulting in liability to you or your company.

COMPUTERIZED PRINT-OUT

As information delivery becomes more sophisticated, you may invest in a personal computer and a program which gives you a standardized report form. Clients may or may not get much benefit from this kind of report, depending on whether it is overly detailed or short and clearly understandable. This depends on how the report's "boilerplate" is designed. You could do it yourself, or purchase a program that lets you fill in the blanks so your printer produces a report which customers see as directed specifically to them.

Using a computer-generated report form has some of the benefits of the BAR, along with some of the benefits of the letter report. It can make phrasing consistent, which can help protect your company from liability created by unclear words and phrases. It can also give you a consistent report structure, increasing the perceived level of professionalism.

On the downside, because the computer-generated report is consistent time after time, it may seem cold and mechanical to the person inserting the information. And the precision you want to place a problem in context may be difficult if you are limited to the same approach on every occasion. Your delivery time is no shorter than with a letter report generated by a typist.

In Option 1 above, you must have verbal skills, but the structure of the report is maintained for you and your work product is delivered immediately. This improves cash flow.

In Option 2, you can use the HomeTech format and dictate your report in the car immediately after leaving the site. You must be able to immediately dictate findings as any other busy executive would, using precise language and a concise style. You can collect for your work on-site or at the time the finished report is delivered. There is a high perceived level of professionalism.

Option 3 requires careful notation immediately after making the inspection. Then either you or your secretary will need to glean the information out of the dictated tape and insert it in the proper places when the computer requests it. You will need someone familiar with computer word processing, and you'll need the computer as well as portable dictating equipment.

In the future, more companies may rely on the third option for inspection reports. Your computer can generate typed copy faster than your stenographer in many cases, but the machine will not be able to program every possible reporting situation. You'll still find it necessary occasionally to make a full narrative report. Sometimes a cover letter may be necessary to put the right emphasis on the case.

While you're training in the field to become a seasoned home inspector, the HomeTech BAR is your best assurance of a report that is fast and understandable. A written report should always be available to your client within 48 hours of the inspection, so until your office routine has gotten up and running, the BAR report delivered on-site offers the best all-around service.

Perhaps the single most important rule in preparing a home inspection report is to complete it before your next inspection. It is impossible to remember everything, no matter how good your notes are, unless you do it this way. If you must schedule your inspections further apart at first, do it. In addition to greater accuracy and prompt delivery, the one-at-a-time rule preserves one of the major benefits of this business: "When you're done, you're done." Your paperwork is complete, and you don't have to spend hours in the office at night finishing reports.

Standard Phrases

It is the home inspector's job to educate prospective buyers about all aspects of a house, giving them the information needed to make a buy/don't buy decision, and educating them about the features of ownership and upkeep. Many home buyers don't know or understand very much about single family houses. You are their first practical instructor, and as such your service will be a valuable experience they will remember forever.

If you use standard phrases to talk about a house, you'll communicate more clearly and consistently. Your reporting will be understood by everyone, and you'll be providing a thoughtful, top quality service. Here are some standard phrases you'll find helpful during inspection conferences and in your written reports.

Equipment

As you are talking about appliances, furnaces, water heaters, plumbing systems, windows, roofs, and even electrical wiring, you might say: "We often say that houses last forever if they are structurally sound, but integral parts wear out on a predictable basis. In this house the age is such that many of the integral parts are nearing the end of their normal life cycle."

This language will help buyers to understand that they must face upkeep and replacement of integral parts and that this is something all homeowners expect. When you list the expenses likely in the next one to five years, they will be prepared and not intimidated. They will understand that all houses have this kind of on-going expense.

The X + Y Equation

In discussing a house, especially if remodeling is being considered, use the following statement:

"When you buy a house you need to go through what we call the X plus Y equation. X represents the price you pay for the house today, and Y comprises the expenses you will be likely to incur in the coming one to five years. The combination of these two will tell you the true cost of the house."

This is most useful when the client is planning to work on the house and needs to know the cost prior to making a decision on the purchase, or when the house has substantial defects or maintenance requirements.

Water Problems

When discussing water problems, use this wording:

"Inadequate control of surface water causes 80-90% of water problems. This house sits high enough, and our inspection indicates that a high water table is unlikely, so the wet basement appears to be the result of a lack of roof and surface water control."

Then you discuss steps to correct the problem such as aligning gutters, extending downspouts, regrading next to the house foundation, and so on.

Another statement that creates a dramatic image in the client's mind is this:

"A house is not a boat."

This means it's not possible to build a house that floats in a sea of water, and if water collects next to the house it will come in -- through cracks or through the slab. There is no way you can waterproof a house like a boat.

Plumbing Systems

Before 1935 all houses in America had galvanized pipes, which rust from the inside out. One way to describe this condition is:

"Old galvanized pipes are like hardening of the arteries. As the rust builds up in the pipes, the opening through which the water flows becomes smaller. The rust drops through the vertical pipes and comes to rest in the horizontals, so the horizontal pipes will clog first. If you don't want to replace all the water supply pipes at once, replacement can be a two-step process with the horizontal pipes being changed first. This will step up water pressure substantially. Then five or ten years later, or when the bathroom is remodeled, the vertical pipes can be replaced."

Metal Roofs

Metal roofs that leak are a common source of legal problems for inspection companies. A roof report should contain disclaimers like the following statement:

"Metal roofs will last a long time if kept painted or coated, but if they rust through they cannot be satisfactorily patched and must be replaced. When a metal roof has been coated with tar it is impossible to determine the condition of the metal under the tar."

This clearly explains that if the metal roof is old and coated with tar, you cannot determine the condition of the metal underneath and the roof might have to be replaced. If any spot tarring has been done, the roof will almost certainly have to be replaced. A spot-tarred metal roof will certainly need replacement within six months, and failure to identify this condition can easily leave the home inspector in line to pay for a new metal roof when the old one goes.

Slate Roofs

Bangor or Pennsylvania slate roofs will last between 30 and 50 years and then begin to spall, while Vermont or Buckingham slate roofs will last 75 to 100 years. Most home buyers believe a slate roof will last forever, so it is important to educate them with wording such as:

"There are slate roofs and there are slate roofs. Vermont slate roofs can last 75 to 100 years or more. Although they sometimes have problems with rusted nails and loose shingles, they are durable. Nails or individual slates may require periodic replacement.

"Pennsylvania slates, by contrast, have a normal lifetime of 30 to 50 years, at which time they shale and spall. Once this starts happening, you have an expensive regular maintenance problem. Many home buyers find the best decision at this point is to take off the slate and install solid sheathing and a composition roof."

Slate roofs must be examined and explained in those terms or you will have problems.

Cedar Shingles

Cedar shingle roofs on less than a four- or five-in-12 pitch have a normal life cycle of 12 to 15 years, at which time they rot. This has to be explained in just those terms, as follows:

"Some cedar shake roofs are durable and, depending on the pitch of the roof and how they are installed, can last 20 to 30 years. Cedar shingles, particularly those on a very low pitch roof, have a normal life cycle of 12 to 15 years, when they rot out and must be replaced."

Asphalt and Fiberglass Shingles

Asphalt and fiberglass shingle roofs have a life span of 15 to 20 years, which you should explain as follows:

"Asphalt and fiberglass shingles last 15 to 20 years and you can almost wind your watch by this. If a house is more than 20 years old and has the original roof, the chances are very high that the roof needs replacing. If the house is 10 years old, you probably will need to replace the roof in the next five to ten years."

Change the numbers in this explanation for the southern United States, where asphalt and fiberglass shingles last only about 8 to 12 years.

Aluminum Wiring

Aluminum wiring is often a problem in houses built between 1960 and 1973 because improper connectors were installed. One study showed that a house with aluminum lower branch wiring had a 55% greater chance of an electrical fire than one with all-copper wiring. A bill was introduced in Congress which would have required that aluminum wiring be removed and replaced with copper in existing homes. While this was over-reacting, aluminum wiring does represent a higher risk and some clients will not buy a house without copper wiring. For those people not aware of the difference, use words similar to these:

"Aluminum wiring was used extensively in houses built between 1960 and 1973, and it was routinely installed with improper connectors. Aluminum has a lower coefficient of expansion than copper. The aluminum wire will heat and cool, become brittle and oxidize, and this has caused some house fires. Aluminum lower branch wiring has been outlawed in many jurisdictions and is no longer being installed. You should be aware of these differences.

"In a house with aluminum wiring your options can be pig-tailing the wiring with copper at the switches and fixtures, which has not been a satisfactory solution; or applying a spray to prevent oxidation; or using a new device called Copalum. This treatment costs about $15 per outlet and makes an airtight fitting that attaches copper to aluminum. It has proven to be quite effective."

You might also reassure the prospective buyer that aluminum wiring does not necessarily make the house a bad buy. If there is flickering at any outlet, however, the system should be checked by a licensed electrician.

Identification of aluminum wiring is one of the most important things to look for when you remove a panel box cover. Failure to point out the problem could end up costing you for the repairs.

Heat Pump Systems

The heat pump is a relatively new feature, but is gaining popularity, especially in milder climates. Many clients may not realize the advantages and disadvantages of such a system. Use the following terms to explain:

"Heat pumps remove latent heat from the air and use it to heat a house during the winter. Once the temperature drops below freezing, it is impossible to get enough heat from the air, so back-up heat must kick in. In the past this has been electric heat strips, which operate like a standard electric furnace. This is expensive, about 35% more expensive than the heat pump so that in very cold climates it may not be an energy-efficient system. Some heat pumps use gas or oil as the back-up.

"You should be aware that with a heat pump, air temperature coming out of the ducts is 90-95 degrees as compared to 130-140 degrees with gas or oil. For this reason, if you turn off the thermostat or turn it down at the end of the day, it will take about an hour to bring up the house temperature and the ducts will seem to be putting out cold air. If you adjust the thermostat up more than three to five degrees at a time, the system will revert to the backup heat. Most homeowners with heat pumps keep their thermostats on a steady setting.

"Heat pumps should not be installed in a house more than 15-20 years old that didn't have air conditioning, as there can be problems with the sizing of the ducts. Individual rooms should not be closed off, and all registers should be left open."

Plaster Ceilings

Before 1935, wood lath with plaster was used for interior walls and ceilings in frame houses. Ceilings with this finish are almost always wavy, but this is a cosmetic problem, not a structural one. The explanation should go like this:

"This house has wood lath and plaster. Over a period of time, the wood lath loses its resiliency and pulls away from the frame slightly, causing waves in the plaster surface. This is cosmetic and not structural. The solution is to tear out the plaster and install drywall, or install 1/2" drywall over the plaster using 2-1/2" nails or screws through the plaster into the ceiling joists."

There is usually no problem if the ceiling is solid. Be sure to point out that ceilings which have been papered over plaster can become a major heartburn if the do-it-yourselfer tries to remove the paper before painting. The phrasing should go along these lines:

"With a wallpapered wood lath and plaster ceiling your options are either to live with the paper painted and continue to paint, or to install drywall over the plaster. If you try to remove the paper, you should budget for major plaster repair. Also, patches will be smooth while the rest of the ceiling is wavy. There is usually a visible contrast."

It is also important to point out that cracks come back in plaster walls and ceilings; they don't disappear with minor cosmetic treatment. If a crack is cut out, pointed up, sanded and repaired, it will reappear in a couple of years because it is a symptom of movement. A recent innovation is to use fiberglass tape over existing cracks. You should explain this clearly, or buyers will call back, concerned that cracks mean hidden structural problems which you failed to point out.

Steel Casement Windows

Steel framed casement windows were used commonly from 1945 to 1955, and they are the worst window you can find. While this is a defect, it should not be magnified, and it shouldn't necessarily prevent people from buying a house. A good approach goes like this:

"Nobody loves casement windows. They get sprung, they leak air around the frames, and the cranks get stripped."

You then go on to point out that the buyer can install replacement windows, and the cost can vary widely.

Appraisal

When you are asked to give an opinion on the dollar value of the house, that is the question an appraiser should answer. It is out of your field; you are not an expert, and you're not being paid to give that opinion. If you try to answer the question, you will scare every real estate agent away from recommending your services. Here is what you should say:

"We have enough trouble keeping up-to-date on construction and renovating costs, which is our field of expertise. We cannot keep up on the price of houses. We are not appraisers, and we cannot offer expert testimony on the value of the house."

You might possibly observe that the house is "in the ballpark," especially in the presence of a third party from out of town. You might say the house "does not surprise me," but this is not a highly desirable phrasing.

Unwanted Company

Another situation that often arises is when the sellers want to tag along during an inspection. It is far better not to have the seller present during your inspection. Here is your best retort:

"It is up to you, but I don't recommend it."

You're working for the purchasers. If you can, it's best to express to the seller in a nice way that it's time to back off.

Just the Facts

Do not worry about the effect your findings will have on the purchasers. It is your job to give them the facts, and let them make the decision on buying the house. You will find that you cannot predict what the effect will be.

Inspections have been done on houses where more than $30,000 of work was required to put the house in liveable condition. The roof was bad, the basement was wet, there was no attic insulation, the gutters needed replacing and the kitchen and baths needed remodeling. The inspector was sure the deal would be killed when the facts were presented. But the purchasers said, "That's about what we figured," and bought the house.

On the other hand a house might be in mint condition, and the inspector is thinking, "I'm not sure I can find anything to justify my fee." While reporting to the purchaser, the inspector mentions that there are 3 inches of insulation in the attic rather than the optimum 9 inches for that part of the country. The purchaser immediately insists that the seller must pay for the additional insulation, and the deal falls through because the seller and purchaser cannot agree.

Sample Inspection Reports

The following are actual inspection reports, one narrative and one using the HomeTech Building Analysis Report (BAR) form.

Dear Mr. and Mrs. _____:

Here is our inspection report on the property at _____ . This report is based on a visual inspection of the accessible areas of the building and its integral components. There are no warranties expressed or implied included with this report.

STRUCTURAL: Structurally, the building appeared to be in sound condition. This a wood frame building constructed on a solid poured concrete foundation with aluminum siding on the exterior which we believe to be approximately 15 years old. T-111 siding has been installed on the gable end and this is beginning to show signs of warping. While there were no visible signs of any active wood boring insect infestation noted on the property at time of inspection, we recommend you obtain a termite certificate at settlement and maintain a termite proofing warranty on the property on an annual basis.

ELECTRICAL: Electricity is provided to this house by an underground 200 amp circuit breaker service. Lower branch wiring is copper. Circuitry throughout the interior of the house is Romex non-metallic sheath cable. There is a 100 amp subfeed which is run to the basement for the electric resistance furnace. Duplex outlets and switches tested during the course of the inspection were found to be in operating condition. We would recommend the installation of a smoke detector in the basement, and smoke detectors throughout the house should be functioning at settlement and checked every two to three weeks.

HEATING, VENTILATION, AIR CONDITIONING: Heat is provided to this house by an electric resistance furnace. This is the original installation and should have a life expectancy of 20 to 30 years. The most common problem with these furnaces is that elements have a tendency to burn out but these can be easily replaced. The furnace did heat up when turned up by means of the thermostat. It is important the furnace filter be changed every 30 to 60 days during the heating and cooling seasons to provide proper air circulation throughout the house and also to keep down dust. The system is equipped with an evaporative type humidifier. This should be cleaned prior to each heating season with a vinegar and water solution to prevent the buildup of mineral deposits on the float valve which could otherwise result in seepage to the interior of the furnace. This was functioning at time of inspection. The humidistat control is located on the return air duct. Central air conditioning was also functioning at time of inspection. This is provided by a nominal 2-ton compressor which we believe to be approximately 6 years old. Normal life expectancy of a unit of this type is 8 to 14 years. Both systems need to be cleaned, serviced and adjusted by a competent heating and cooling technician. Air distribution throughout the house does appear to be adequate.

PLUMBING: Water is supplied to the house by a copper supply piping system. The main water supply line and shutoff valve is located in the right front corner of the basement. During the winter months make sure the outside faucets are turned off which can be done by means of the valves located beneath the kitchen sink. Leave the outside faucets open to allow any water standing in the pipes to drain to prevent them from freezing. Drain, vent and waste lines are Schedule 40 PVC and appear to be in good condition. Hot water is provided to the house by a 50 gallon electric water heater. This unit is only about 3 years old and may still be under warranty. A unit this size will provide ample hot water for up to 5 people and has a life expectancy of 8 to 12 years. There is a rusted through J trap beneath the kitchen sink and this will need to be replaced. The hall bath fan should be hooked up and operating. You have to make sure the area where the tub meets the tile and the tub meets the floor is kept caulked with a good rubberized silicone sealer. This will prevent water from seeping in behind the walls and on the floors with resultant damage to the ceilings below. The subflooring next to the commode in the master bedroom bath is soft and evidently there had been some water causing deterioration of the plywood. You may wish to have this replaced in the future. Water pressure throughout the house does appear to be adequate.

BASEMENT: The basement of this house was dry at time of inspection with little sign of any previous water penetration. There is a solid poured concrete foundation. In order to maintain a dry basement it is necessary to control exterior surface water and we will elaborate on this in the "Exterior" paragraph. There is a sump pump in this basement which was dry at time of inspection. The main bearing for the house is a built up wood beam which does appear to be in sound condition as do the floor joists running off it. When you hook up your washer and dryer make sure the dryer is kept vented to the exterior of the house to prevent excessive moisture from building in the basement. This basement would be a good candidate for renovation.

KITCHEN: The kitchen appliances were functioning at time of inspection. These do appear to be the original appliances. Statistically, cooking and refrigeration equipment last 15 to 20 years, while dishwashers and disposers have a normal life expectancy of approximately 5 to 12 years. The exhaust fan is a ventless type fan and the charcoal filter should be changed at least once a year. Wood cabinetry has been screwed to the stud walls and appears to be in sound condition. Sheet goods have been installed over the subflooring.

GENERAL INTERIOR: Walls and ceilings of the house are done in drywall and in sound condition throughout. Windows are aluminum sliding type windows which are equipped with storms. The storm windows should prove to be cost effective so far as heating and cooling bills are concerned. The double glazed sliding glass door should also be fuel efficient, but there is a broken seal on the pane. New wall-to-wall carpeting has been installed over the subflooring and this is in good condition. In the attic area there is approximately 8 to 10 inches of fiberglass insulation which does bring the house up to contemporary energy conservation standards. Ventilation is provided by attic louvers and should be adequate. There is a solid masonry party wall between the individual townhouse units.

GENERAL EXTERIOR: The roof on this house is an asphalt shingle roof approximately 2 to 3 years old. The normal life expectancy for a roof of this type is 15 to 18 years. The present roof has been installed over the existing roof shingles. The roof was observed from the ground using binoculars. It is important to make sure the flashing around the vent collars and along the parapet walls is kept intact so moisture cannot gain access to the interior of the house from these areas. The gutters do appear properly aligned but the downspouts should be extended well away from the side of the house so water cannot pond against the foundation walls. Grading on the exterior should be maintained so there is a positive pitch of approximately 1 inch per 1 foot of run extending 5 or 6 feet away from the side of the house. This, too, prevents water from pooling along the walls which causes saturation of the soil and eventual seepage into the basement. The exterior trim needs to be caulked and painted. The T-111 siding is beginning to curl at the left side of the gable. There are birds in the powder room fan and they should be removed.

All in all the house appears to be in sound structural condition with adequate maintenance. We would not anticipate any major expenses being incurred with ownership of this property other than routine maintenance which can run anywhere between $500 and $800 per year.

Should you have any further questions concerning the property, please do not hesitate to call.

Sincerely,

Sample Inspection Report

YOUR COMPANY NAME
Business Address
City, State, Zipcode
Telephone Number

BUILDING ANALYSIS REPORT

Name _J. Watson_
Customer

Address _123 Main St._

City _Our Town_

State, Zip Code _State_

Property Location
456 Second St.
Our Town

This is our report of a visual inspection of the readily accessible areas of this building, in accordance with the terms and conditions contained in the PRE-INSPECTION AGREEMENT, which is a part of this Report and incorporated herein. Please read the REMARKS printed on each page and call us for an explanation of any aspect of this Report, written or printed, which you do not fully understand.

Date of inspection _5/27/90_ Weather conditions _rain_ Outside Temperature _65°_

PRE-INSPECTION AGREEMENT
(PLEASE READ CAREFULLY)

COMPANY agrees to conduct an inspection for the purpose of informing the CUSTOMER of major deficiencies in the condition of the property. The inspection and report are performed and prepared for the sole, confidential and exclusive use and possession of the CUSTOMER. The written report will include the following only:

- structural condition and basement
- electrical, plumbing, hot water heater, heating and air conditioning
- quality, condition and life expectancy of major systems
- general interior, including ceilings, walls, floors, windows, insulation and ventilation
- kitchen and appliances
- general exterior, including roof, gutter, chimney, drainage grading

It is understood and agreed that this inspection will be of readily accessible areas of the building and is limited to visual observations of apparent conditions existing at the time of the inspection only. Latent and concealed defects and deficiencies are excluded from the inspection; equipment, items and systems will not be dismantled.

Maintenance and other items may be discussed, but they are not a part of our inspection. The report is not a compliance inspection or certification for past or present governmental codes or regulations of any kind.

The inspection and report do not address and are not intended to address the possible presence of or danger from any potentially harmful substances and environmental hazards including but not limited to radon gas, lead paint, asbestos, urea-formaldehyde, toxic or flammable chemicals and water and airborne hazards. Also excluded are inspections of and report on swimming pools, wells, septic systems, security systems, central vacuum systems, water softeners, sprinkler systems, fire and safety equipment and the presence or absence of rodents, termites and other insects.

The parties agree that the COMPANY, and its employees and agents, assume no liability or responsibility for the cost of repairing or replacing any unreported defects or deficiencies, either current or arising in the future, or for any property damage, consequential damage or bodily injury of any nature. THE INSPECTION AND REPORT ARE NOT INTENDED OR TO BE USED AS A GUARANTEE OR WARRANTY, EXPRESSED OR IMPLIED, REGARDING THE ADEQUACY, PERFORMANCE OR CONDITION OF ANY INSPECTED STRUCTURE, ITEM OR SYSTEM. COMPANY IS NOT AN INSURER OF ANY INSPECTED CONDITIONS.

It is understood and agreed that should COMPANY and/or its agents or employees be found liable for any loss or damages resulting from a failure to perform any of its obligations, including but not limited to negligence, breach of contract, or otherwise, then the liability of COMPANY and/or its agents or employees, shall be limited to a sum equal to the amount of the fee paid by the CUSTOMER for the Inspection and Report.

Acceptance and understanding of this agreement are hereby acknowledged:

H. Inspector _5/27/90_   _J. Watson_ _5/27/90_
Company Representative Date   Customer Date

PAYMENT RECORD

Total Fee $ _180.00_   Paid by: ☒ Check ☐ Cash ☐ VISA ☐ MasterCard ☐ American Express

Account No._____ Name on Card _____ Exp. Date_____

Company Representative _H. Inspector_   Date _5/27_ 19 _90_

46 HomeTech Form 403 B.A.R.

STRUCTURAL

Type of Building	☒ Single ☐ Duplex ☐ Rowhouse/Townhouse ☐ Condo ☐ Other _____ ☒ Gable Roof ☐ Shed ☐ Hip ☐ Gambrel ☐ Mansard ☐ Flat ☐ Other_____ *Likely 50 or more years of age*
Structure	☒ Wood Frame ☐ Brick and Block ☐ Solid Brick ☐ Other _____ Foundation Wall: ☐ Poured Concrete ☒ Block ☐ Brick and Block ☐ Solid Brick ☐ Other Floor Framing: *Main 1st floor support steel I beam & columns, 2x10 joists, diagonal subfloor, good overall condition. Good joist size for span* Wall Framing: _____ _____ Roof Framing: _____ _____ ☒ Signs of Water or insect damage ☐ Extensive ☐ None noted ☒ No major structural defects noted—in normal condition for its age *Past termite damage at front wall, new plate & partial treatment*

BASEMENT (OR LOWER LEVEL)

Basement	☒ Full ☐ None ☒ Open Walls ☐ Closed ☒ Open Ceiling ☐ Closed ☒ Extensive present basement storage, visibility limited
Floor	☒ Concrete ☐ Dirt ☐ Other _____ ☐ Satisfactory ☐ N/A
Floor Drain	*Near rear door* ☐ Satisfactory ☐ N/A
Sump Pump	☐ Operating ☐ Not operating ☐ French Drain ☒ N/A
Basement Dampness	☒ Some signs ☐ Extensive ☐ None noted ☒ Past ☐ Present ☐ Not known *Some minor stains, but no problem during rainfall at time of inspection. Stains at base of wall at storage room.*
Crawl Space	☐ Readily accessible ☐ Not readily accessible ☐ Satisfactory ☐ Conditions observed ☐ Method _____ ☒ N/A ☐ Conditions not observed Floor: ☐ Concrete ☐ Dirt ☐ Other _____ Clearance below joists: ☐ Ample ☐ Inadequate Dampness: ☐ Some signs ☐ Extensive ☐ None noted ☐ Vapor barrier ☐ Insulation ☐ Ventilation

HEATING AND COOLING

Heating System	Fuel: ☒ Gas ☐ Forced Air Furnace ☐ Oil (see remarks on page 4) ☐ Electric ☒ Gravity Hot Water Boiler ☐ Forced Hot Water Boiler ☐ Steam Boiler ☐ Radiant Heat ☐ Electric Baseboard *oversize for house as* ☐ Heat Pump *numerous radiators removed* Capacity _144,000 BTU_ Age: _40-50, original_ When turned on by thermostat ☐ Fired ☐ Did not fire ☐ Satisfactory ☐ N/A *Nearing end of normal service in terms of efficiency.* *System should have automatic water feed & water* *supply to boiler should be left open.* *Boiler needs cleaning. Adjust aquastat & check on* *gas shut-off valve — frozen.*
Fuel Supply	☐ Oil tank in basement ☐ Buried ☐ Other _____ ☒ Public gas supply ☐ Tank ☐ Other _____
Heat Exchanger	☐ Checked ☐ Not checked ☐ N/A ☐ Have condition checked before settlement *Boiler appears to be wrapped in asbestos. When boiler* *replaced, be sure removed along EPA guidelines.*
Distribution *Bleed radiators*	☒ Radiators ☐ Galvanized pipes ☐ Copper Pipes ☐ Satisfactory ☐ Pipes not visible ☐ Convectors ☐ Baseboard Convectors ☐ N/A ☐ Ductwork *would help to add circulating pump* *Rust at main exhaust, vent stack satisfactory* *w/ water heater & boiler running together*
Humidifier	Type: ☐ Atomizer ☐ Evaporator ☐ Steam ☐ Satisfactory ☒ N/A
Filter	☒ Washable ☐ Disposable ☐ Electronic ☐ Satisfactory *not accessible* ☐ N/A
Supplementary Heat	Location Type *Electric furnace w/ heat pump* ☐ Satisfactory _____ _____ ☐ Satisfactory _____ _____ ☐ Satisfactory
Cooling *Not installed electrically*	☒ Central Air ☐ Room Units ☐ Gas Chiller ☒ Electric Compressor ☐ Heat Pump ☐ Satisfactory Compressor Age: _new_ Capacity: _2½ ton_ ☐ N/A ☐ Tested ☐ Not tested (see Remarks on Page 4) ☐ Ductwork *not tested — at top floor left* *bedroom the supply & return are adjacent,* *too close together*

PLUMBING

Water Service	☒ Public ☐ Private (See Remarks on page 6) Pipes: ☐ Copper ☒ Galvanized ☐ Brass ☐ Plastic ☐ Lead ☐ Unknown *Some rust on incoming line. Eventually budget to replace, possibly within 5 years.*	☐ Satisfactory ☐ N/A
Interior Pipes	☒ Copper ☐ Galvanized ☐ Brass ☐ Plastic ☐ Unknown Water Pressure: ☒ Tested ☐ Not tested *New bath copper, old bath combination copper & galvanized. Pressure a bit low, likely due to incoming galvanized service pipe (aerator removed)*	☐ Satisfactory ☐ N/A
Hose Bibbs	☐ Operating ☐ Not operating ☐ Not tested *Disconnected at front*	
Waste Discharge	Waste Pipes: ☐ Copper ☒ Galvanized ☐ Brass ☒ Plastic ☐ Lead ☒ Cast iron ☐ Unknown ☐ Slow drain *Vent pipes plastic & galvanized* *No leaks noted*	
Waste Disposal	☒ Public ☐ Private Septic System (See Remarks on page 6) ☐ Not known	
Hot Water Heater	☒ Gas ☐ Electric ☐ Oil ☐ Integral w/heating system Capacity ___30___ Gal. Ample for ___3___ people Age: _10-11_ ☐ Pressure relief valve and extension *Remove insulation on hot water line (it's melting from gas vent pipe exhaust). Budget to replace water heater, sounds worn, w/gurgling noise.*	☐ Satisfactory ☐ N/A

BATHROOMS

Fixtures	*1) Bathtub at 1st floor is not back vented other than tied to main house stack* *2) Cannot verify if new bath turned original stack to wet vent or if done properly. County rough-in inspection would verify.*	☐ Satisfactory ☐ N/A
Bathtub	☒ Built in ☐ Leg tub ☐ Access Panel ☐ Fiberglas Surround *Whirlpool not tested (not wired)*	☐ Satisfactory ☐ N/A
Stall Shower	☐ Ceramic Tile ☒ Fiberglas ☐ Metal	☒ Satisfactory ☐ N/A
Ceramic Tile	☒ In mastic ☐ In mortar bed	☒ Satisfactory ☐ N/A
Plumbing Leaks	☐ Some signs ☒ None noted	
Ventilation	☐ Fan ☒ Window	☒ Satisfactory ☐ N/A

ELECTRICAL

Service	Service line entrance: ☒ Overhead ☐ Underground Service cable size: _____100_____ Amps Type: _____Alum._____ ☐ Satisfactory Panel Box _____200_____ Amps ☐ Fuses ☒ Circuit Breakers *upgrade* ☒ Grounded *Service not complete, will not need upgrade* *size cable. Entry cable doesn't appear to have proper drip loop*	
Circuits	Quantity: ☐ Ample ☐ Satisfactory ☐ Ground fault interrupters *Add several circuits* *Only 6 at present – 5 needed for kitchen alone*	
Conductors	Branch wiring: ☒ Copper ☐ Aluminum ☐ Satisfactory	
Outlets and Fixtures	☒ Random Testing *Many not completed* ☐ Satisfactory ☐ Smoke Detectors *Many rooms have only one outlet* *Raise light at first floor closet*	

KITCHEN AND APPLIANCES

Cabinets	*New installation not complete* *Be sure to get all warranty papers*	☐ Satisfactory ☐ N/A
Countertops		☐ Satisfactory ☐ N/A
Dishwasher	☐ Operated ☒ Not Operated ☐ Air Gap Age: _____	☐ Satisfactory ☐ N/A
Disposal	☐ Operated ☒ Not Operated ☐ HP _____ Age: _____	☐ Satisfactory ☐ N/A
Range/Oven	☐ Operated ☒ Not Operated ☐ Gas ☐ Electric Age: _____	☐ Satisfactory ☐ N/A
Refrigerator	☐ Operating ☒ Not operating ☐ Frost free ☐ Ice Maker Age: _____	☐ Satisfactory ☐ N/A
Other Appliances		☐ Satisfactory ☒ N/A
Floor	☐ Resilient Tile ☐ Sheet Goods ☒ Other *Quarry tile*	☒ Satisfactory
Ventilation	☐ Exhaust fan ☐ Ductless ☐ Vented to outside *Suggest fan vented to exterior*	☐ Satisfactory ☒ N/A
Clothes Washer	☐ Operated ☒ Not operated	
Clothes Dryer	☐ Operated ☐ Not operated ☐ Gas ☐ Electric *N/A* ☐ Vented	

INTERIOR

Floors	☐ Hardwood ☐ Softwood ☐ Plywood ☐ Wall-to-Wall Carpet ☒ Satisfactory ☐ Not visible *Floor sloped at breakfast room, previously a porch* *Dip in floor (living to dining) likely due to weight of partition* *Minor high spot at first floor bedroom not a problem*
Walls	Plaster on: ☐ gypsum lath ☒ wood lath ☐ masonry ☐ Satisfactory ☐ Drywall ☐ Wood *Some ceilings covered with drywall*
Ceilings	Plaster on: ☐ gypsum lath ☒ wood lath ☐ Satisfactory ☒ Drywall ☐ Wood
Stairs	*Need to stiffen basement stairs* ☐ Satisfactory *Need railing at top floor for safety* ☐ N/A *, Repair*
Fireplace	☒ Flue liner ☒ Damper ☐ Operated ☐ Not operated ☐ Satisfactory ☐ Metal Pre-Fab ☐ Clearance _____ " ☐ Wood Insert Stove ☐ N/A ☐ Clean before use *Does not appear to have fire brick.* *Likely will need mason to make operable.*
Doors (inside)	*Some doors removed , otherwise* ☒ Satisfactory
Windows	☒ Double hung ☐ Casement ☐ Awning ☐ Sliding ☐ Satisfactory ☒ Wood ☐ Metal ☐ Vinyl ☒ Insulated glass— *rear 1st* ☐ N/A ☒ Storm Windows — *most have ;* *floor* *caulk them?*
ATTIC	
Access	☐ Stairs ☐ Pulldown ☐ Scuttlehole ☒ No access ☐ Satisfactory ☐ N/A
Moisture Stains	☐ Some signs ☐ Extensive ☐ None noted
Storage	☒ Heavy ☐ Light ☐ Not floored *In closets of bedrooms*
Insulation	Type: *Fiberglass* Ave. Inches *6¼* ☐ Satisfactory Installed in: ☒ Rafters ☐ Floor ☐ R Rating *19* ☐ N/A *At visible area near furnace*
Ventilation	☐ Window ☐ Attic Fan ☐ Thru house fan ☐ Louvers ☐ Satisfactory ☒ Ridge Vent ☐ Soffit Vent ☐ N/A *Suggest adding soffit vents — ridge alone not adequate*

51

ROOFING

Roof Covering	Location _House_ _Garage extension_ _____ _____ _____	Materials _Asphalt - 2 shingles not flat_ _Tin_ _____ _____ _____	Age _new_ _____ _____ _____ _____	* ☒ Satisfactory ☒ Satisfactory ☐ Satisfactory ☐ Satisfactory ☐ Satisfactory

How observed: _From ground with binoculars_
* Need 1/2 shingles outside front left bedroom windows, then flash like adjoining
One rafter rotted at end (front porch), minor rot at 2 in rear

Roof Leaks	☐ Some signs ☐ Extensive ☒ None noted

| Flashing | ☒ Galvanized ☐ Copper _Need to paint at chimney_ ☐ Satisfactory ☐ N/A
At valleys aluminum extends only 4-5" — should be 8" or more |
|---|---|

| Gutters and Downspouts | ☒ Aluminum ☒ Galvanized ☐ Copper ☐ Vinyl ☐ Wood ☐ Satisfactory ☐ N/A
Will eventually get galvanic reaction between 2 metals. Extend leader at rear & front — water is pooling |
|---|---|

EXTERIOR

Exterior Doors	_Repair garage door operation_	☒ Satisfactory

Windows and Skylights	Roof windows and skylights: ☐ Moisture Stains ☐ Extensive	☒ Satisfactory ☒ None noted

Exterior Wall Covering	Location _Front and sides_ _Side and rear_	Materials _Wood clapboard, recently painted_ _Cedar shingles_ _Beaded fir & Texture 1-11_ _____	☒ Satisfactory ☐ Satisfactory ☐ Satisfactory ☒ Satisfactory ☐ Satisfactory

| Exterior Trim | Soffit, Fascia, Eaves: ☐ Signs of rot ☐ None noted
Some minor repairs, good overall | ☐ Satisfactory ☐ N/A |
|---|---|---|

Chimney	☐ Brick ☐ Metal	☒ Satisfactory ☐ N/A

| Garage/Carport | ☒ Garage ☐ Carport ☒ Attached ☐ Detached
☐ Door operator ☒ Operated ☐ Safety Stop
Door hardware needs repair/replacement. Settlement crack at midpoint & at end where garage extended — seal with mortar | ☐ Satisfactory ☒ N/A |
|---|---|---|

Porch	Floor: ☐ Wood ☐ Concrete ☐ Other _____	☒ Satisfactory ☐ N/A

GROUNDS

Grading	General grading, slope and drainage: *Nearly flat* *Best to extend downspouts 10' or to street, due to flat site. 2' tiers*	☐ Satisfactory ☐ N/A
	Grading and slope at house wall (within 5 feet from building): *Need to build up grade & extend downspouts* *Slope should be 1" per foot for 5-6' from foundation*	☐ Satisfactory ☐ N/A
Sidewalk and Walkway	☒ Concrete ☐ Brick ☐ Flagstone ☐ Other _____	☒ Satisfactory ☐ N/A
Driveway	☒ Concrete ☐ Asphalt ☐ Other _____ *Asphalt at end. Remove asphalt or install new drive for complete drainage — less expensive to remove existing and repair*	☐ Satisfactory ☐ N/A
Window Wells		☐ Satisfactory ☒ N/A
Retaining Wall	☐ Brick ☒ Block ☐ Stone ☐ Other _____ ☒ Mortared joints ☐ Dry ☐ Weep Holes *Minor lean at right side* *No repair needed at present*	☐ Satisfactory ☐ N/A
Trees and Shrubbery		☒ Satisfactory ☐ N/A
Fencing	*Needs paint — galvanized*	☐ Satisfactory ☐ N/A
Wood Deck	☐ Signs of rot ☐ Extensive ☐ None noted	☐ Satisfactory ☒ N/A
Patio, Terrace	☐ Concrete ☐ Brick ☐ Slate ☐ Other _____	☐ Satisfactory ☒ N/A
Steps to Building		☒ Satisfactory ☐ N/A
Outbuildings	☐ Not observed	☐ Satisfactory ☒ N/A

SUMMARY

List of items not operating or with major deficiencies: *or needing repair*

Have contractor check sizing of boiler. Service clean & adjust boiler.
Bleed radiators.
Fireplace needs firebrick at inner hearth & walls to be operably safe
Electrical service & system not complete. Heat pump not
installed electrically.

List of some important items requiring possible repair or replacement within next 5 years:

Item	Estimated Cost Range
Boiler (asbestos removal extra)	2700 - 3300
Water heater, 40 gallon	400 - 500
New water service	approx. 1500
Grading and downspout extensions	600 - 800
Ventilation	approx. 500

Miscellaneous minor repairs and expenses during the first year of occupancy are estimated to be

$_____ to $_____.

Remarks
A return visit would be necessary to check final electrical & heat pump connections
Check on contractor getting proper county permits and inspections for electrical, plumbing and mechanical systems.

3. TOOLS OF THE TRADE

Promises You Can Keep

There are any number of advantages to the home inspection business. If you like people and houses, providing a valued service and having the variety of different houses and clients every day keeps you interested. You're an expert professional, a source of valuable information that can influence monumental decisions. Your customers love you: they constantly say things like, "My, you know a lot about houses!" As a result, your psychic income is at least as high as your monetary income.

You make promises you can keep, unlike other construction-related fields where the whims of subcontractors and demands of the day can put circumstances beyond your control. As a home inspector you have only to be when and where you say you will be. Nobody is likely to derail your appointed rounds.

Being in control goes a step further, if you manage your time carefully and use advanced techniques. When you're done, you're done. At the end of the day you gather your report copies, or drop off tapes at the office, and the information is conveyed. Your job's complete and you can avoid paperwork. You're almost entirely in control of your own work processes and are no longer at the mercy of "shoppers" in the marketplace.

There is good money to be made in the field. In larger metropolitan areas where inspections are assumed to be part of buying a house, almost every inspector who works full time can make $50,000. Some subcontract inspectors earn in the seventies. There is an enormous and expanding market in information, and as a home inspector you have a solid piece of the future.

The skills required are relatively easily mastered. The tools used to make convincing reports are personal qualities and attitudes. Delivery of the information requires talent, which is a product of knowledge and attitude.

On the other hand, there are mechanical tools that you'll need to perform your inspections and managing them correctly will help you to master the personal skills.

Mechanical Tools

Your major home inspection tools are a sharp eye and a keen mind, coupled with a good sense of what to say and when to say it. Using the report forms and strategies outlined elsewhere will make the delivery of your product uniform in quality and appearance. Tools, in the sense of hardware used to help you do your job of data collection and processing, will include most if not all of the following items.

Binoculars. Any moderately good pair will do. Use these for checking condition of the roof ridge, shingles, guttering, and first visual inspection of trim or any other items that you can't readily walk up to. Make the first check from street front, and walk around to each side of the house to examine the areas.

Folding 12- to 14-foot ladder. This ladder may be required so you can climb up on a flat roof, or to look into attic access areas. The best solution is a type of folding ladder that can accordion down to fit in the trunk of your car. You don't want to present the appearance of a tradesman or home improvement contractor, but a professional. It would be better to drive a sedan or clean, late-model station wagon than a pickup truck in most urban situations, although this doesn't necessarily hold true in every section of the country.

Powerful flashlight. This is an essential tool, and you should use either a marine-type light with long-life batteries or one of the new pocket lights with very bright, adjustable beam. You will need to use this repeatedly, from inspecting under sinks for leaks to reading model numbers or water capacities of hot water heaters, to looking in attic spaces and crawl spaces under the house. Always carry a good quality flashlight. It is as important as your clipboard.

Steel tape or other measuring device. This is handy for describing the precise locations of cracks, stains, etc.

Screwdriver, flat blade and Phillips. This can be a convertible type and can be pocket size. The major use is to remove electrical panel box covers, inspection panels at bath tubs, and possibly equipment panel covers in the heating/air conditioning system. While you don't need machinist's quality tools, a good strong screwdriver will look good and will last indefinitely.

Pliers, slip-joint. You may not use these on every house you inspect, and may leave them in the car along with the ladder. Occasionally you will need to unfasten clamps, etc., and it is good practice to have the tools you need for any eventuality.

Receptacle tester. This tool will give you a reading on the correct wiring of three-prong electrical receptacles throughout the house. If they're wired correctly, one combination of lights will so indicate. If there's a faulty ground, a bad hot lead, or several other questionable conditions present, you'll be able to identify the problem with the tester. Newer models also test GFI outlets. It is a tool that provides valuable information, is durable, and looks professional. It creates an appearance of high-tech investigation without taking you out of the generalist's camp.

Thermometer. Use this to test cooling of air conditioners and refrigerators, if you wish to go so far.

Cassette tape recorder. This tool stays in your car most of the time, but is invaluable when you're clearing out information on one inspection before arriving at the next appointment. You'll have notes, taped report, and a preliminary list report all neatly categorized within 30 minutes of completing the inspection which is what you need to have your report completed before your next inspection.

Clipboard and pens. Your clipboard should have a folding plastic cover with paper pouch to carry extra forms that you won't want on the writing surface during the inspection process. The cover serves a dual purpose. You can use it as a knee protector on wet grass when you have to inspect very low crawlspace access doors. The folding cover is inexpensive, but it is a professional's tool and suggests that you are adept at collecting and directing information.

Moisture meter. This is becoming standard equipment for many home inspectors because it is so helpful in detecting whether wet spots are active or not.

Specialized technical testing equipment. This collection will include items not considered necessary in the generalist's home inspection approach, such as electrical amp probe, microwave tester, gas tester, CO tester for heat exchangers and other items.

Cellular car telephone. If you are working in an area with cellular telephone service, this is practically a must. If you are caught in traffic or running late, or some other problem comes up, the cellular phone can be a real lifesaver.

Using the Tools of The Trade

To establish a perspective on the use of the tools you'll need, a short inspection scenario is in order. Assuming you either have a good area map or know the area very well around the site, you should observe as you approach the house. Look for major arterial streets, major utilities lines, power transformer location, manhole covers in the street, water meter locations, gas and electric meters, all visible from the street or from a fast walk around the property.

Stop some distance from the house and use the binoculars to sight along the roof ridge, down the eaves, along the tops of windows, along porches that may or may not be sagging, and at each window and door. This long-distance visual examination should reveal data on the house including general structural soundness, foundation settlement, decay at the eaves, condition of the electrical weatherhead, chimney and flashing, disintegration around exposed materials in the porch, possible differential settlement, condition of window sashes and presence or absence of storm windows. Make notes to yourself at this point.

You should have a good idea of the type of utility services provided, whether the house is on a public sewer or has its own septic tank, whether leaks are likely to be discovered around specific areas of the roof, if there has been dramatic shifting of the foundation, and in general, how well kept the house is and whether it has asbestos shingle siding. The combination of these data points will begin to suggest to you the general age and condition of the house.

If you discover large cracks in masonry foundation or brick veneer, the measuring tape will help you accurately describe the size and location of cracks. It is better to report a 3/8" wide crack than to call attention to a "large crack". More notes.

Having crossed the front porch and opened the front door, look down at the floor. Let your eye travel across the floor in both directions to spot bows, sags, warped boards, stains and burn marks. Note anything that is outstanding. Look around the room, at all visible wall surfaces and electrical outlets, and then look up. Notice how the wall/ceiling joint is carried out, whether with crown molding or other trim, and if there are separations, sags, etc.

You're looking for obvious structural and mechanical/electrical problems, and if you know the code well enough, you may spot code violations. Beginning in this room you may have use for your flashlight, receptacle tester, screwdriver, steel tape, clipboard and note pad.

If the house has a basement, this will be your first major stop. Check the condition of the floor, walls, presence of floor drains, sump pump, crawl space (unexcavated area), and look for evidence of dampness or water stains. In the basement and crawl space you may need to check minimum dimensions from floor to bottom edge of joists above, as an indicator of the difficulty of converting the basement into a habitable room. You'll need to know code requirements for minimum ceiling height, ingress/egress and ventilation. Information you bring with you will be applied to the data collected in order to produce a reasonable analysis of the house.

In the basement you'll find the heating/cooling equipment and probably the water heater, water conditioners, humidifier, air cleaners, and other mechanical equipment. Any test equipment relating to these items will come into play now, and you might have a small pouch to keep all the testers neatly organized when you're moving from site to site. More notes.

As you move through the house you'll need a screwdriver to take the cover off the electrical panel box and possibly to remove inspection plates on other equipment; a receptacle tester to check out visible electrical receptacles (if they're the three-prong type); your tape, flashlight, screwdriver or pocket knife, and your report forms.

After making the first trip through, bring in your ladder if needed to inspect the attic. Make a mental inventory at this point of additional tools you'll need and return those you have no further need of to the car when you walk out. By establishing a specific system for using and returning your tools, you can avoid misplacing equipment and appear well organized to others present. The second pass through should be one in which you collect remaining data and any equipment which you left in place for a test, such as thermometers, etc.

You'll not only create an impression of good organization and a controlled professional analysis, you'll actually develop the good habits that separate the true professional from the haphazard amateur.

Tools are important in carrying out your inspection. Style and organization are almost as important as the tools. You can actually inspect a house with just a few tools if you know the signs and are a good detective. Then if something appears to need further study, you can have every feasible item neatly organized in your car and quickly available.

You will need a definite approach to the inspection process. Here is one sequence found to be successful.

✓ Outside walk around

✓ Begin in basement (If crawl space is muddy, save this for last, and don't enter a crawl space if you suspect electrical problems in the house.)

✓ Kitchen, including electrical panel box

✓ Interior

✓ Attic/roof

✓ Outside walk around

Go through this cycle twice, except that you go into crawl spaces and attics only once. When you are done, you will have collected all the information typically needed on a "structural inspection using standard, visual techniques."

Instructions for House Inspectors

Any well-managed organization must have guidelines so that members of the team will be able to measure their behavior against a norm. This creates an understanding of what is expected, and helps accomplish the objectives of the company. Job responsibilities and expectations are established by the owner or a personnel executive or committee. When expectations are clear, it will be obvious that the organization needs each member and no one is being "picked on" by unreasonable requirements.

HomeTech uses a set of Instructions for House Inspectors. While each inspector may not be held to every item, the following guidelines apply in most cases.

DRESS: All inspectors will dress in a neat manner. For men this means wash-and-wear slacks, dress shirt with tie, rubber soled shoes shined. Sport coats or suit coats will be worn October 1 through May 1, and are optional from May through September.

CARS: Cars or light pickups should be late model, within 3-5 years old, and show no visible defects (dents, rust-through), loud mufflers, or unusual decoration. Maintain a prosperous, professional appearance at all times and in all aspects.

EQUIPMENT: Inspectors should carry the following equipment:

- √ Good flashlight
- √ Carpenter's ruler or steel tape
- √ Clipboard
- √ Combination Phillips/regular screwdriver
- √ 14' fold-up ladder
- √ Area maps for all areas in which work is done
- √ Portable tape recorder with extra blank tapes
- √ Different types of inspection reports: BAR, On-site
- √ Receptacle tester
- √ Thermometers
- √ Moisture meter

Inspectors should not chew gum, smoke cigarettes or pipes, or eat lunch while waiting for or during an appointment, whether or not the real estate agent or the client is present.

All appointments must be kept. The inspector should be prompt unless unusual circumstances warrant. (These relate only to delay caused by previous appointments, NOT personal business. If the inspector will be more than 15 minutes late, a courtesy call must be made informing the client of the situation.

If no one is at the appointed address to admit the inspector, the following procedure will be observed:

A. The inspector will wait in the car or at the premises until 15 minutes after the appointed time.

B. After 15 minutes, the inspector will call the office and/or realtor's office to determine if there has been any reason announced for the delay.

C. After making the call, the inspector will return to the appointed location and wait until 30 minutes at least after the appointed time. If no one shows by then, a card is left in the door and the inspector will be free to leave for the next appointment. Staying longer than the stipulated 30 minutes is at the inspector's discretion. (If the inspector has a car telephone, a call should be made immediately to the office, purchaser, or agent. Information obtained from that call may make waiting the 30 minutes unnecessary.)

REPORTS: All reports will be dictated or completed as soon as the inspection is finished and before another inspection is undertaken. If additional information is necessary from outside sources, a draft report will be dictated. Reports (or dictated tapes) must be delivered to the office by noon on the first business day following the inspection (Saturday and Sunday appointments report by noon on Monday). If a BAR report is used, it must be filled out and delivered to the client on site at the inspection; otherwise, a narrative report must be used.

Inspectors can promise 24-hour preparation of narrative reports unless exceptional circumstances interfere, such as weekends or clerical overload at the office. This privilege should not be abused. Only cases in which a report is urgently needed to release a contingency should be given priority treatment. Routinely, a 48-hour office turnaround is the maximum time allowed, except over weekends. Inspectors are responsible for delivery of their own reports. Reports may be faxed to the customer. No inspector should require another inspector to deliver a report for any reason.

When reports are promised on a priority basis of less than 48 hours, they should be dictated on a separate cassette and placed in the rush box in the office. Put a note with the cassette specifying the time promised and the delivery method.

All cassettes placed in the office box for typing must be clearly labelled with the name of the inspector and the client. This labelling should be on the same side of the tape as the report.

Where callbacks are necessary to check such items as air conditioning, plumbing, etc., the inspector should insist that the client or agent be responsible for setting up the time and gaining access to the property. Only in exceptional instances should the inspector assume responsibility for scheduling a return appointment.

When roof reports are to be done after the original inspection, the report should be specified as incomplete, and a separate note should be left in the roof report box in the office. The note left in the roof box must contain the following information:

    A. Name, address, and telephone number of client
    B. Type of roof
    C. Approximate height of roof above ground
    D. Time necessary for report to be completed
    E. Specific items to be checked: gutters, flues, etc.
    F. Name of inspector

When special roof inspections must be hired from outside, the additional cost of the roof inspection fee will be shared equally between the inspector and HomeTech.

When inspection reports are to be sent to parties other than the original client, specific permission should be obtained from the client. If this permission is not obtained, make a note for the office to call and get the permission. The inspector should obtain the proper address and name of the party to receive the report.

All inspectors will be paid on fees actually collected through Monday on the week preceding payment date. This payment will be for 40% of the total fee except in special situations, and excepting travel time. The entire travel time charge will be paid to the inspector. (Rate of payment varies by individual, depending on time in service and rank in the organization.)

TRAVEL TIME: Travel time will be charged for inspections over 30 minutes from the Beltway, except in those areas where we are pursuing additional business aggressively: Frederick, Manassas, Woodbridge, Annapolis, Baltimore and Westminster. Inspections across the Chesapeake Bay or in West Virginia will be charged travel time.

If an inspector has made a presentation to a real estate agent and wants the office to send follow-up letters and brochures, put the agent's card in the box marked for this purpose in the office. Follow-up materials will be sent within one week.

SPECIAL FEES: If realtors request an inspection on property for their own personal use, the inspector has the option of deducting $25 or more from the fee. This promotional price reduction is at the discretion of the individual inspector.

AVAILABILITY: Inspectors are expected to be available for appointments from 7 AM to 8 PM five days a week, and from 7 AM to 5 PM on Saturday. Sundays are optional, unless there is a need for an inspection to meet a contingency deadline.

At least 24-hour, and preferably 72-hour, notice should be given to the office for prior commitments which limit availability. All inspectors should make themselves available for as much time as possible.

Once an inspection appointment is confirmed, no change in the time should be made unless absolutely necessary. HomeTech wishes to avoid a reputation of changing appointment times to meet our needs, and wishes to meet the client's timetable in every way possible.

Inspectors are independent subcontractors, responsible for all their own expenses. The ONLY expenses paid by the office are as follows:

✓ Cost of entertaining directly relating to soliciting business

✓ All costs of office administration and reports

✓ Cost of office space, telephone and utilities used on the premises of HomeTech

✓ Business cards and stationery, brochures, and other promotional printed matter

✓ One-half the cost of an inspector's cellular telephone, including leasing and access fee

Inspectors must pay their own expenses for:

✓ Insurance
✓ Worker's compensation
✓ Car expenses
✓ Personal telephone calls from the field
✓ Equipment (tools, tape recorders, etc.)

Inspectors are responsible for returning customers' calls as necessary.

Inspectors are expected to keep abreast of industry trends and new products, including updating on current prices and general techniques.

Inspectors meetings will be held on a weekly basis and it is the individual's responsibility to attend these meetings.

Tips for Inspectors

Whether you are operating a one-person show with a desk and filing cabinet in a corner of the den, and a telephone answering machine for those times you can't catch the call, or are subcontract inspecting on a full time basis for a larger inspection company, there are a number of tips to make the job easier. Here are a few.

When making an appointment for an inspection, take the name, current mailing address and telephone number (two or more if possible) of the client or prospective purchaser.

If your fee scale is related to the selling price of the property, be certain to learn the selling price of the house on the first call. Fees range from $150 upwards, often increasing $1 per thousand for sales prices over $150,000. This is a simple and easily understood fee schedule. Clients will understand that more expensive and complex homes take more of the inspector's skill and attention -- and may create greater liability for the inspector.

Get the name of the real estate agent(s) involved so you will know ahead of time whether the property is familiar to you. You might also ask whether the house is single family or multi-family, frame or masonry construction, and the approximate age of the residence. Getting the agent's name not only ensures you have the name for your marketing database, it allows you to address the agent by name when you arrive at the inspection.

Have up-to-date maps of the entire area in which you might do inspections, including outlying areas within a 50-mile radius.

Even though you have good maps, be sure to get directions to the property from the person placing the inspection order. Maps may be inaccurate, as well as directions. If you have trouble finding the place, use common sense and try reversing turns and transposing numbers that you have written down.

Always dress professionally, but don't wear expensive business suits to do a home inspection. Wash-and-wear clothes are best, with a washable raincoat in winter. You can keep a pair of coveralls (disposable is fine) in the car to use when you must get into attics and crawl spaces. Be sure to arrive at the inspection in professional dress, and get the coveralls only when you need them.

It is important to appear as a real estate and construction consultant and dress accordingly. While a sport coat and tie may not be necessary in the country, all professionals in a large urban environment will be dressing according to the local custom, and you should, too!

Remember how important it is to establish your image as a peer of the client. You're making an analysis that can affect the outcome of a major commitment, and your demeanor should be friendly but also crisp and professional.

When you are inspecting a house that is vacant or no one is at home, be sure to ring the doorbell and also give a voice warning before entering the premises. You don't want to startle someone who may be asleep or working quietly in the basement. Give another voice warning when emerging from the basement if there are other people in the house, to avoid any surprise at your sudden appearance.

Always check the house twice from basement to roof. This allows you to double-check problems and pick up details the second time around. It also gives you a head start in case the client isn't there at the time scheduled to begin the inspection. Nobody is likely to object, but if they do, you can point out that your process requires two trips through the house and the client or real estate agent is welcome to follow you on the second round. If you sense that the client has expected you to spend more time on one part of the inspection, possibly in the basement, explain that you go through twice. This should relieve any fears and assure people that you are doing a thorough job.

If a third party is accompanying the client, make a special effort to bring that person into the conversation by asking their opinion on some minor item. This is a way to deflate tension and avoid the adversarial relationship that might otherwise develop.

You should determine immediately whether the client and real estate agent have mutual and complete confidence. If the purchaser asks you not to talk in the presence of the agent, honor this request. If the agent stays within earshot when you are ready to give your final report, ask to speak to the client for a moment in private. Find out whether there's any objection to the agent's presence. If there is, remember who you are working for and follow the client's wishes. Don't ever consider releasing a copy of the inspection report to the agent without the client's prior approval.

Ask the client at the start of the inspection whether there are any specific questions. Many people do have special concerns, which will give you a clue to pay special attention to certain items. Bring the client into the actual inspection wherever possible. Some people don't want to be involved, but others want to learn everything they can, especially people buying their first house. These are the best clients for you to give a lesson in household care and maintenance along with a thorough inspection tour. Most inspectors agree that this is the best kind of client, because while it may slow you down a little, the perceived value of your service is going to be very high and the potential for claims against you is reduced dramatically.

Some of the things you should point out as you lead the prospective homeowner through the house:

✓ Locate the main water cutoff to the street.

✓ Locate the electrical main panel box, ta~ has fuses or circuit breakers.

✓ Locate the water line to the hur ~d off seasonally.

✓ Locate gas meter, electric m~

✓ Show whether clothes dr~ ~ow it can be if it's not.

✓ If washer and drye~ ~ryer hookup is and explain how to co~

✓ Locate separa~ ~ate which receptacles are dedicated fo~ ~rcuits.

✓ Locate s~ ~s there is a potential moisture proble~

If y~ ~sement, show the client but withhold your opinio~ ~ly and then re-inspect the basement. Only then~ ~ an active problem, whether it is caused by un~ ~be corrected easily.

Detecting ~ ~use

As soon as po~ ~ing the inspection, you should determine the age of the structure. There are va~~s clues all around you. One of the most common is the imprint on the underside of the toilet tank cover. If it's the original toilet, this imprint will give the date of manufacture, which most likely will be within one calendar year of the date of construction. In older homes, where there is a high probability that the bathroom has been remodeled, you must be aware of other clues that date the home. Typical characteristics are listed below.

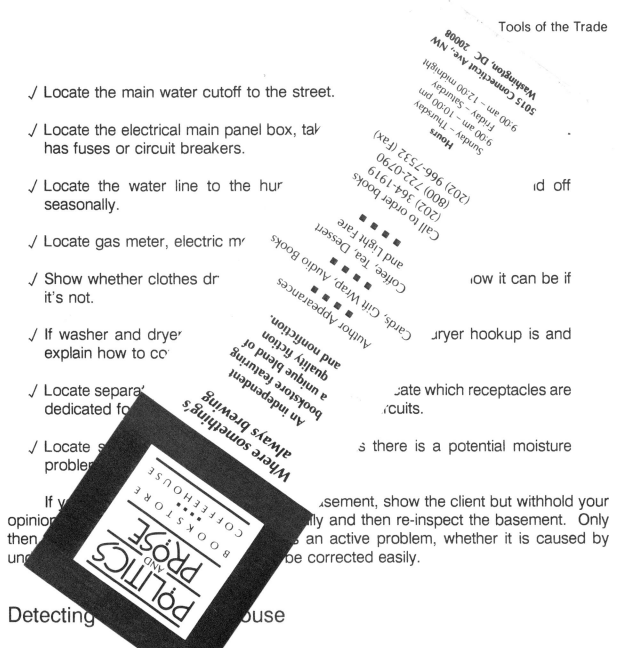

Before 1935, exterior walls of masonry buildings were usually plaster directly on brick. Furring strips became common after 1935, with a one-inch air space between plaster and brick.

Before 1935, wood lath and plaster were the most common materials for interior walls and ceilings. After 1935 and almost universally by World War II, rock lath covered with plaster became the standard interior wall finish.

In the early 1950s, drywall became the predominant material for walls and ceilings, and it became almost universal by 1960. Only custom-built homes today use "hard coat" finish of rock lath and plaster.

Steel casement windows were common through the early 1950s, and were predominant from 1945 to 1954.

Double hung windows with sash cords and weights were used from the turn of the century until about 1950. Since then, balance frames have been used instead of sash cords.

Coal was the common furnace fuel before 1925. Most hot water boilers from that era were converted to oil, and may still be in use.

Galvanized water pipes were used almost exclusively until 1935, and in all but the better custom homes through World War II. Copper pipes came into general use at the end of the war and became a standard in most homes by the mid-1950s.

Pine floors were very common before 1930, in many cases used in the upstairs and family living quarters, with oak used in the formal reception rooms.

Most homes built before 1930 had leg tubs in the baths, which almost always indicates lead waste plumbing over to the main cast iron stack.

By knowing symptoms of decay and key signals that tell you the age of a house, you will develop into a "house doctor with a good bedside manner." You'll begin to develop your own techniques with experience, and as you develop a personal style, your professional polish will take on an individual style. In truth, very shortly, you will become your own highly competent professional.

4. THE REAL ESTATE WORLD

The Hand That Feeds Us

Here's your self-employed home inspector, crossing town after a full day of successful inspections. He places two cassette tapes containing four completely dictated reports in the box for the secretarial service and switches on the message machine.

"Howard, this is Amelia Topseller of Prestige Properties. We need an inspection on the old Hornbarrier property and we need it by 5 PM tomorrow at the latest. Please give me a call at home tonight."

Howard flinches. Meeting Amelia's request will mean rescheduling the next morning's work, but it must be done. This makes him uncomfortable, but he doesn't feel he has any other realistic choice. Since Howard went into business, Amelia has ordered more inspections from him than any other single customer in the past six months.

He looks on her account as a true "bread and butter" relationship. Sometimes he thinks that because Amelia represents the "bread" in this business arrangement, she expects Howard to butter her up with superior service on extremely short notice. So far, circumstances have been favorable and none of Howard's inspection reports has caused any of Amelia's deals to fall through.

Looking toward the day when a family uses his report to escape a contract, Howard worries a little. He thinks of her as a friend, but he isn't confident of the direction the business relationship may take once Amelia feels the sting of a canceled contract on his account.

Such a situation is repeated around the country where a closeness has developed between real estate companies and a preferred inspector or company. The result is an uncomfortable symbiosis, unless clear lines are drawn and understood at the outset. From Texas to California to Pennsylvania, as home inspections are done on the majority of properties that change hands, the real estate industry is fostering development of the home inspection business. These liaisons can develop into mature business relationships, or they can end up as short encounters where nobody achieves a long-term benefit.

The home inspector runs the risk of being too much "in the pocket" of the companies that provide his bread and butter work. Sometimes the relationship is unnecessarily sharp and rocky. By laying the proper groundwork, you can avoid both of these uncomfortable situations.

After studying this chapter you will have a better knowledge of what motivates real estate people, what you can expect from them -- and why. You'll also be better prepared to foster a relationship that will bring in business for your company while giving the real estate agent good support and professional service.

The Best Times for the Industry

Moderate market conditions may be the best for home inspections. In an overheated housing market, with buyers competing for everything available and bidding prices up thousands of dollars over the original asking price, properties change hands "as-is", with the buyer assuming the risk of paying for any necessary repairs in order to get the house.

On the opposite extreme is the seriously ailing real estate market. Here sales have all but disappeared, and the agent is looking at a net income of perhaps $15,000 this year. The salesperson who is desperate for that commission check will avoid any chance of interference. In the first case, the momentum of the market can cast the home inspector as unnecessary baggage; in the latter, any possible impediment to a sale is to be avoided at nearly all costs.

It is most often the real estate salesperson who calls for a home inspection. If the real estate industry isn't exactly in control of the home inspection market, it does at least have a very major impact. Home inspectors will find it to their great benefit to understand the dynamics of the real estate transaction in order to carve out their own niche in it, form viable alliances, and preserve their professional integrity in the bargain.

How a Real Estate Agent Views the World

It's a common mistake to view a real estate salesperson as a non-creative broker of the goods and services of others: the "real producers" who build, landscape, design, remodel, own and buy and sell houses. In fact, the agent works diligently and must manage a host of professional tasks in order to earn the commission check. And even after the check has been deposited there may be additional services for good agents to deliver. These are usually in the form of helping new families adjust to the environment, make contacts in the community for schools, churches, civic clubs, and other organizations.

Real estate agents must be unquestionably professional, both in appearance and manner as well as knowledge of their business. They must study for state examinations which require knowledge of real estate law, liability, consumer practices, salesmanship and marketing, appraisals, finance, VA/FHA regulations and requirements, as well as other subjects including pre-qualifying potential customers. They must know how to deal ethically with the public and provide the level of service the industry itself demands.

Good agents will be well informed about a wide range of available housing in the area, writing copy for property advertising, the psychology of selling and dealing with sellers, types of construction, quality of construction, community services and utilities, taxes, and a variety of other subjects. Agents who deal primarily in single family detached housing will have a different set of concepts than those who work in the inner city with row houses, restoration properties, condominiums and multi-family buildings. And real estate professionals must pass state examinations and be licensed by the state before they can legally transact business.

New-home salespeople have one marketing expertise, while those who concentrate primarily on pre-owned houses have another. Commercial real estate is an entirely different arena. Brokering businesses and selling unimproved development land and farms is another. Yet another area of real estate specialty is land development, turning countryside into subdivision, arranging for zoning, utilities, design and survey work, grading and paving, lot clearing, special marketing promotions and general lot sales.

Since most real estate agents and brokers have a college education, those who don't must deal on an equal level. The real estate market is primarily white collar. Thus the representative of a successful firm will speak and write articulately.   This is a requirement to generate the confidence of the buying and selling public. The fact that many real estate professionals migrate into the field from other areas such as education, reveals more about the stress and rewards of the schoolroom than it does about the entry requirements and rewards of the real estate profession.

Real Estate Commissions

When a property goes on the market, the real estate broker writes a contract with the seller stipulating a percentage of the price as the sales commission. This varies depending on the kind of property. It is also negotiable between seller and broker. As a general guideline, residential single family property will bring the agency a 6% commission. Unimproved property, depending on size and value, may bring anywhere from 6% to 10% for subdivision lots. Commercial real estate may bring from 3% to 10% sales commission to the listing agency.

Real estate brokers want an exclusive listing for their agency. This means the agency is paid full commission on the property no matter who sells it. If a property sells under an exclusive listing and generates a 6% commission, the internal split might be: 1.8% for listing agent, 1.8% for selling agent, and 2.4% for the agency. On an $80,000 home sold by a real estate agent other than the one who wrote the listing, the selling agent would receive a total commission of about $1,440.

If the same property were placed in multi-listing and a salesperson from another agency made the sale at $80,000, the commission would be less for the selling agent. Multi-listings traditionally net a 50-50 split between listing and selling agency. The salesperson might receive a 50-50 split with the agency, meaning the salesperson's gross commission on the property would amount to no more than $1,200.

Agencies often make a split with their salespeople in the following manner: listing agent, 30%; selling agent, 30%; agency, 40%. Therefore, agents who sell a home for which they wrote the original listing, may earn either 50% or 60% of the gross commission. At 50%, the listing agent could stand to gain as much as $2,400 on the $80,000 sale.

This is a before-tax amount, used only to illustrate the point that listing and sales agents must work diligently to earn a substantial income. Also, they must have the benefit of favorable market conditions and a certain amount of good luck.

Unlike home inspectors, real estate agents have no guarantee that their work will result in a paycheck. If the sale falls through, time and effort have been expended with no visible reward. It is easy to see how some agents -- particularly those feeling financial pressure -- would fear anyone's imposing on the deal and perhaps killing it.

Anxiety Relief

Your main message as a home inspector must alleviate the anxiety felt by a sales agent. You have two powerful tools to meet that objective.

It is the condition of the property and not the inspection report that will be the determining factor, in nine out of ten cases where a sale falls through.

Far better the defects be discovered so the buyer can make a decision based on full knowledge than for major defects to be discovered later. The agent might be liable for a greater repair bill than the commission ever amounted to.

The Liability Tightrope

Real estate people are facing an additional layer of risk that requires even greater specific knowledge: Latent liability. The hidden, concealed, or not discernible structural flaws in a property can come back months or years after the sale, perhaps seriously threatening an agent and seller who represent a property as sound. One dictionary of real estate defines **liability** as, "The opposite of asset."

Fortunes have been made in real estate. But the honest, professional agent is always suspended on a tightrope between a possibly fraudulent seller, a shady developer, or a disreputable builder, and the general public. The income is not regular, but the bank interest payments are, and the family must keep on eating in times of slow sales as well as boom times. Some studies have shown that the average real estate professional in the country may earn less than $19,000 annually. Yet the knowledge required to succeed in the business is not reduced for low earnings. Neither is the liability.

Understandably, the industry will go to great lengths to protect its professionals and educate them on ways to serve the public better while protecting their hard-earned assets. This is the context of the industry that calls up to order a home inspection from your company. You can do three things for real estate professionals, from their viewpoint:

√ You can provide an expert analysis which will lessen the potential for future surprises.

√ You can increase the professional quality of the package being assembled, along with appraisals, surveys, credit reports, legal and financial documentation. Thus you can bolster the buyer's confidence in the whole process.

√ You can kill the sale.

For the real estate agent, it's a balance of benefit and risk. When an inspection report does kill a sale, both the agent and the inspector had better be prepared to continue the business relationship. Otherwise, everyone loses: buyer, seller, sales agent, listing agent or broker, attorneys, bankers, home inspectors, and the economy at large. It is not the lost sale that causes the damage, but the rip in the fabric of a professional business community.

Hurt feelings are annoying, but a failure to deliver the best possible professional service for families creates a black eye on the community. Better that the real estate agent be made to realize the client's interests have been protected, and the small matter of one lost sale can be overcome by finding another house for the buyers. But a major defect in a house that goes unreported is a time bomb waiting to explode and ruin agent and inspector alike.

You must educate the local real estate industry that you are not a deal-killer, but a measure of valuable protection and reinforcement for the transaction. In more than three cases out of four, the deal will go through. Whether it does or not, everyone involved will be better off if the house has been inspected professionally.

The Effects of Consumerism

In today's economy, consumerism is a growing concern which has led to greater sophistication in the real estate community and contributed directly to the emergence of the home inspection. Lenders are required to issue truth-in-lending statements and to delay contract closings for a specified "cooling off" period. More detail than ever is required in the real estate arena, and the national oversupply of lawyers is contributing to an increase in litigation because every professional must earn a living.

The era of consumer protection is also the age of civil lawsuits, as people rush to clarify financial rights and obligations before the courts. The home inspection industry has grown parallel to this legal development. Openness and trust in the marketplace defer to technical experts who are hired to define and advocate the interests of the client/consumer.

Houses are more complex than they were a generation ago, and the outdated attitude of "let the buyer beware" has been replaced by a web of ongoing obligations for sellers. This reflects an effort to protect the consumer as well as a standardization of expectations. In an economy driven by large corporations where regional or nationwide job transfers are common, we have become a nation of strangers.

Neighborhoods no longer connote any particular sense of community. The houses they live in represent the major financial investments of families. Houses bought and resold in a few years -- hopefully at a significant profit -- are no longer really "homes" in the sense of family roots so much as they are major holdings. In these temporary residences, marketability outweighs personality just as property condition and location are more important than love of place.

The tendency to migrate was felt strongly in the Sun Belt during the early 1980s when factory closings in the north sent thousands of workers to warmer climes, where there were jobs to be found. Housing demand all but outpaced available supplies in some areas. Because buyer protection was a hot issue, legislatures heeded the voice of the real estate industry and Texas, now regarded as the most litigious state in the nation, became the first state to require licensing of home inspectors. There the realities are ironic, in that the Texas tradition of doing business face-to-face on just a handshake has collapsed under the weight of consumerism. The first state requiring licensing will by no means be the only one. State licensing of home inspectors is on the horizon -- for a number of reasons.

Legislative direction, placing liability for latent defects in structures, has given the real estate industry reason to support an alliance with the home inspection business. Legal liability for any structural defects that might become obvious only a year or more after a property is sold can create quite a threat to both seller and real estate agent. If a portion of that liability can be controlled by placing the responsibility on home inspectors, then the real estate industry has a clear motive for encouraging inspections. While this might constitute an uneasy alliance, based more on the real estate sales-person's perception of negative factors than on actual shared goals, the effect is one that home inspectors can turn to their advantage.

It simply means inspectors had better do a good job, and be prepared to face the music whatever the outcome.

Eyes on Texas

And what outcome can be expected? As part of the Texas licensing law for home inspectors, the state has established a system of appeals and formalized hearings. Inspectors must take a written exam, show evidence of having completed a 90-hour classroom course, and make personal disclosures. Each year, license renewal and exam fees are charged and a $250 payment must be made into a state recovery fund, which is not an insurance pool but a fund to pay the administrative costs of commission hearings.

When a dispute arises between an inspection company and a home buyer, and it is not resolved before coming to the state review board, all parties are subpoenaed and must appear in Austin for a hearing. Since a negligent inspection is a violation of Texas law, the inspector who loses a case before the state appeal board can suffer loss of license and be required to pay damages.

The inspector can also be required to pay for repairs the inspection may have failed to catch.

Horror stories continue to be told because there is no compiled body of knowledge and no Standards of Practice. According to one Texas state official, "What's hot about inspections today is, inspectors are getting sued."

The fees charged for an inspection range from $85 to around $250, and among 1,300 licensed inspectors in the state the range in performance is great indeed. The same state official described inspection reports ranging from a 25-page, word-processed, single-spaced report down to a written note, "Congratulations, your house passed!"

While tensions existed initially between real estate and home inspection industries, the Deceptive Trade Practices Act has resulted in a sellers' Property Condition Disclosure Statement, an attempt to move the liability for known defects away from the real estate agent's door with these words:

"This is not a substitute for any inspections or warranties the buyer(s) may wish to obtain. The following are representations made by the owner(s) based on the owner's knowledge and are not the representations of the listing Realtor, any cooperating broker, and their agents, the _____ Board of Realtors, or the Board's Multiple Listing Service."

The disclosure form is a checklist of several dozen items from structural components to environmental hazards, and condition of equipment and systems. It requires the seller and the listing agent to think about most of the items that would be covered by an inspection. While no actual inspection is required, the process makes the seller and selling agent aware of the value of a home inspection.

The real estate industry is moving to absolve its practitioners of liability resulting from known or latent defects. On the Texas Association of Realtors (TAR) residential listing agreement, a paragraph on Property Defects states that all known defects, including latent structural defects, are revealed on the attached Disclosure Statement. The form goes on:

> The Texas Real Estate License Act states, "Latent structural defects and any other defects do not refer to trivial or insignificant defects but refer to those defects that would be a significant factor to a reasonable and prudent purchaser in making a decision to purchase."

The agreement then holds harmless the real estate agents and their representatives for any oversight made by the seller in the disclosure of conditions, and states:

> "Owner is further advised that, if said information is false or inaccurate, Owner may be held liable for damages caused by said falsity or inaccuracy."

Clearly, the Texas Association of Realtors has a system to take the heat off its members to the extent possible. This is the motivation, and not altruism, that has led to the next development: the TAR actively supports the concept and practice of home inspections. TAR officials do not require that members of the association call for an inspection before a property changes ownership, but the vast majority recommend it.

Working with an inspector, the sales agent has an opportunity to mediate between buyer and seller to protect the status of any sales contract. First, is the reduction in liability to the seller and selling agent when a licensed inspector makes a reliable report. Second, the agent can prevent panic when defects are pointed out -- helping to preserve the sale and ensuring that the commission check will indeed be available for deposit at the appointed time.

A consumer information brochure distributed by TAR lists benefits to the consumer and describes types of inspection services. It prepares the buyer for negotiating repairs without breaking the contract, describes the sequence of having repairs made (after loan approval but before closing), and points out the necessary time frame for having the inspection. Selecting an inspector, and remedies in the event of an unsatisfactory inspection are covered as well. The brochure makes a strong recommendation for inspection of both new and pre-owned homes.

National Association of Realtors

At the National Association of Realtors in Washington, DC, officials point out that the NAR's Code of Ethics requires members to reveal all known facts about a building to the buyer, even though many state laws may not require as thorough disclosure as Texas or California. One NAR official feels that having a home inspection is generally good buyer protection. He lists several standards he would recommend in choosing an inspection company:

√ Good referrals

√ Not in the home repair business

√ Provides a written report (no joint reports with other companies like termite inspectors)

√ Invites the buyer to attend the inspection

The emphasis is on major structural condition, leaky roofs or basements, electrical, heating and plumbing systems. Home inspectors who nit-pick cosmetic or small items are not appreciated by the real estate industry for obvious reasons: they kill deals.

Generalist inspections are more easily integrated into the business community, because a balanced approach can identify flaws in a house without overplaying them. Creating a proper context is the area where cooperation between inspectors and real estate agents can be most productive and rewarding.

Every professional realizes the value of associated professions. And each group, including the buyer, has or should have one goal uppermost in mind:

To construct the most complete useful description of all aspects of the impending sale, including legal, financial, physical condition and future costs (taxes, maintenance, repairs) so the buyer gets no surprises down the line.

This goal requires strict adherence to professional ethics as well as technical expertise.

Recognizing their future depends on referrals from satisfied customers, today's real estate professionals must meet the new era of consumerism. Enlightened managers accept the customer's perception of satisfaction as the benchmark for a solid business based on growth. It is becoming commonplace for the professional real estate people to recommend, and practically insist, that the home inspection be part of the real estate transaction.

Understanding: A Two-Way Street

Among the services you can provide to real estate professionals -- in addition to good work and fast turnaround -- are reassurances that you understand their situation and want them to understand yours. You can set up a regular routine of calling on agencies to have a cup of coffee with agents and talk over the market in general. Or you can occasionally take one of your better agent contacts out for a business lunch, and be sure to pick up the check.

This is a deductible business expense, and a courtesy that most people sincerely appreciate. At lunch, discuss common concerns, and offer a little of your perspective on the inspection business, but don't put on a hard sell or ask for any specific return. The business lunch is a communication and social ritual. Show that you have polish and an understanding of the unwritten social graces. It will be appreciated.

Remember that your philosophy is that of a generalist. You have to know as much as any technical inspector about houses and technical matters that can be discovered by visual means. Then you have to go the extra distance by being a good communicator. That means your demeanor will always be positive, you'll never alarm either real estate agent or prospective home buyer.

Balance is the key. You must find strengths as well as weaknesses in every situation. That goes for your inspection presentations in the presence of buyers, and your written report. Knowing enough to talk intelligently about legal and liability issues can be reassuring to the realtor as well.

Keep all real estate agents in your area well supplied with your brochures and business cards. Be sure the agents you are trying to win over know your name and business location, have a few of your business cards, and have seen at least one example of a written report, plus a site report or checklist if you use them.

Regular personal contact is the key. To begin with, ask to talk to the broker or sales manager. Then ask low-key questions such as, "Are you presently using home inspections as part of your sales package?" "Who are you using? Are you happy with them?" "Is there any way I can be of help?" Your goal is to give a talk at a regular sales meeting (see page 87 for a sample presentation).

If you speak at a sales meeting, you should get some inspections almost immediately. If you follow up with a call every month or so, and use direct mail, you can make that office a permanent part of your business. But if they do not see you in person, and get to know you somewhat, your chances of recommendation are slight.

You must be convincing with your real estate agent contacts if you are going to present yourself as a good source for educating the first-time home buyer. Pointing out that the buyer and the agent should attend the inspection is easy to do, and you can give reasons why they should.

But you might also tell the agent you believe clear communication is so important that you will call the buyers, if they can't be present, and offer to discuss any points in the report with them at no additional charge. Present yourself as a resource person with a sincere desire to be helpful by giving proper weight to strengths and weaknesses.

It's also a good idea to offer to appear as a resource person at a local real estate board luncheon, to serve on a panel or a board committee, to present a slide show or a talk about home inspections in general. The more understanding you can create between you and the real estate community, the better business you will generate over the long run.

Remember, since sales agents usually work on a commission-only basis, they do legwork routinely with no specific guarantee of a reward. By offering to provide service for them in the same way they provide service to the purchasing public, you create a common bond. Human nature will tend to make you a member of the group when you pay those psychic and personal dues. In the long run you will be building a stronger base by knitting yourself into the fabric of the business community.

When you sit down over coffee or lunch with a real estate professional, a good idea is to give them one or two personal copies of the Real Estate Data Sheets or the booklet "Inspecting a House for Listing and Sales Purposes" produced by HomeTech. Sign it and include your office phone number written in ink on the cover, with your business card attached. You should review the valuable information the data sheets contain, and know it well, before giving away the material. It is a handy reference for real estate agents and the quality of the information will be appreciated.

Just because you have sold the sales manager or broker on the quality of your company, do not expect to get every inspection from that office. Agents are in business for themselves, and make their own decisions. It is common for an agent to provide three names for a client to choose among, but even so the agent can probably control which inspector is used.

You will probably go through cycles with individual agents. An inspector may do ten inspections in a row for one agent, then kill a deal, and the agent goes to another company. That company may do another ten in a row, then kill a deal, and the agent will come back to the first inspector. Don't be surprised if you do inspections for one agent who disappears for a few months and then reappears.

An agent will have a priority list of inspectors to call. If a house is sold on Monday night and the purchasers want a home inspection, next morning the agent will start calling down the list. The first inspector who can do it at the preferred time will get the job. But there is almost no way an inspector the agent doesn't know, a "pig in a poke", will be asked to inspect a house.

Hidden Benefits of the Home Inspection

As real estate people become accustomed to using home inspections, they discover benefits which weren't apparent. Here are four major benefits to the real estate sales professional:

The prospective home buyer likes the house more after the inspection is made. This is because a good inspector doesn't limit the report to deficiencies in the house: the positive features are discussed as well. The balance a hired expert creates usually encourages the hesitant buyer to be more comfortable with the decision to purchase. The "cold feet" syndrome is not easily addressed by the sales agent, because of inherent conflicts. A realistic assessment made by an authoritative expert usually sets the psychological stage for the big commitment.

The inspector teaches first-time home buyers what to expect once they become homeowners. Many first time home buyers don't know basic home care and mainte-nance. Having relied on landlords in the past, they may have no idea of their responsibil-ities and may tend to overlook normal maintenance once they have moved in.

A good home inspector will cover maintenance if the buyers are present during the inspection -- as they should be. The result is educated owners who are more likely to be satisfied because they know what to do, who to call if repairs are necessary, and whether or not to make the call in the first place. This relieves the pressure sometimes placed on the sales person to "hold hands" with novice homeowners until they are comfortable.

The home inspector helps the buyer make financial plans. Homeowners who have just laid out large amounts of money to purchase are not pleased at surprises that translate into big expenditures once they've taken possession. The inspector tells them about the condition of the house and its integral parts, and gives them a list of anticipated expenses for the next one to five years.

This helps them make an intelligent decision about purchasing the house, and enables them to budget for upcoming expenses. If knowledge is power, this kind of knowledge can empower the buyer to navigate successfully through tough times.

When the real estate agent calls for an inspection, that builds a relationship of trust with the buyer. Real estate professionals dealing with used houses must first sell the client on their services and then develop a trust which grows into loyalty. With that foundation, the agent can keep the house hunters through several promising properties until they find the one that meets the family needs and budget.

For this agent, a home inspection is not a threat. If a contract is submitted and the inspection reveals a major defect or imminent expense, the agent can help negotiate the solution. Either the contract can be withdrawn, the price renegotiated, or the purchaser can decide to go ahead armed with knowledge of the condition. If the decision is made against a particular house, the real estate agent keeps the loyalty, and the client, until a match between house and family can be made. This approach increases the perceived professionalism of the real estate agent in the eyes of the home buyer. The result can be future referrals.

How You Can Help

As state requirements on real estate contracts vary, so do standard forms that are used to write sales contracts. In Texas, the condition of the property is spelled out in great detail. The seller's responsibility to correct defects discovered by the home inspection is covered, and a maximum dollar amount is stipulated in the contract. In other states, the standard form spells out that electrical and mechanical systems shall be "in good working order at closing", with specified remedies for defective items discovered prior to closing. Inspections are at the buyer's expense, and repairs are generally paid for by the seller, unless the seller refuses. In that case the buyer can accept the property in "as-is" condition or terminate the contract and get back the escrow deposit.

Often contracts stipulate that mechanical/electrical appliances can be conveyed un-repaired, without sellers notifying buyers of the condition, and giving the buyers no remedy after closing the sale. The wording makes a careful inspection all the more important. Standard contracts designed by the real estate industry usually set the closing date as the last time sellers can be held liable for electrical or mechanical deficiencies in the house.

You should know the conditions in your own work area. The liability for unknown (or latent) defects may creep up on sellers, real estate professionals, and inspectors alike. It is probable that real estate agents and brokers will be protected by the standard clauses in the contract forms. For example, read the blanket statement found in TAR documents:

"Broker(s) and sales associates have no responsibility or liability for inspec-tions or repairs made pursuant to this contract."

Even though a printed contract may attempt to cut off the seller's responsibility for defects in a house at the time of closing, the courts can decide in the buyer's favor. Essentially, the liability still exists and it will be brought to rest somewhere -- very likely on the home inspector, in part or in full -- no matter what.

Be sure you know what the standard sales contract in your area contains. In some areas the contract says that the house will convey at settlement in the condition it was at contract ratification, without stipulating that electrical, plumbing and mechanical items are in working order. In this situation, you might recommend to the purchasers that they put in that stipulation when they release the contingency, in effect writing in a property condition clause to the contract.

You can offer a service by recommending contract clauses to be included when an offer to purchase is made. HomeTech makes the following three suggestions:

"Contract contingent on inspection by a building expert [within 72 to 96 hours], and report satisfactory to purchaser."

"All well and septic systems to be checked by local health authorities prior to settlement and determined to be in good working order."

"If repair or replacement is necessary, cost to be borne by the seller."

Subject to Inspection by a Building Expert . . .

The first clause clearly sets out the terms of the inspector's order. Now it's easy for the purchasers to recognize that this contract embodies protection for them. The short time frame is necessary to be fair to the seller. Most sellers don't want to take their houses off the market for more than 3-5 days.

It's necessary to resolve the contingency as soon as possible. Also, because the outstanding contingency delays the entire home purchase process, the quicker the home inspection can be made and issues resolved based on the report, the faster financing can be arranged, appraisals scheduled, etc.

This clause also gives the purchaser a safety escape from the contract under almost any circumstance. Using "satisfactory to purchaser" as a catch-all phrase makes the contract work like a simple three or four-day option on the property. The purchasers can retrieve their deposit money by simply stating that the report indicated unsatisfactory items. As a practical matter, for the cost of an inspection alone, the house can be held off the market until the buyer can make a rational decision.

Systems to be Checked by Health Authorities . . .

The second clause is a protection for everyone concerned. The buyer is clearly informed by the wording that the home inspector is not the person to check these systems. While, for example, it is almost a foregone conclusion in many parts of the country that a property with spring-fed water supply will not pass health standards, this wording puts the responsibility for a safe water supply and septic system squarely where it should be -- on the seller.

Substandard water in some states can be grounds for a major judgement against the person conveying the property. By using this wording, the issue is clearly between seller and buyer, and dissociates sales agent and home inspector from any liability in that area. You should also point out to your client that your inspection does not cover septic or private sewage systems, or private water supply.

Cost to be Borne by Seller . . .

The third clause is negotiable. A dollar maximum could be set, or a percentage of the sale price might be agreed upon, but repairs for major items should be covered in the sales contract in order to avoid any complicated negotiations and haggling after defects have been pointed out. A stipulated flat dollar amount can cause a problem if the purchaser says to the inspector, "You better find at least $500 of repairs because the seller has already agreed to that amount." This makes the home inspector into a nitpicker, rather than a professional reporting on conditions.

Negotiable and Non-Negotiable Items

Most contingency clauses can re-open negotiations if the purchaser is not satisfied with the results of an inspection. However, most -- although not all -- standard real estate contracts include a **non-negotiable property condition clause** that reads something like this:

> "The house will convey at settlement in the same condition that it was in at time of contract ratification, normal wear excepted. All electrical, mechanical and plumbing items will be in operating order."

This means, for example, that if the contract was ratified on June 1st and the house was in good condition, but on September 1st (settlement date) the dining room ceiling had fallen down because of a roof leak, then the seller would be required to fix the leak and patch and paint the ceiling. This would be non-negotiable. Likewise, if a number of windows were broken by vandals between contract ratification and settlement, the seller would have to repair them. Neither of these situations qualifies as normal wear.

All electrical switches, plugs and lights must be working. If a cover plate is missing, it must be replaced. If a duplex outlet has reversed polarity, it must be fixed. In some cases, an overloaded main electrical service (such as 30 or 60 amps) must be increased.

There can be no plumbing leaks, the hot water heater must work, the drains must all drain normally, and the faucets must not be in need of washers. A defective shower pan must be replaced, and usually any defective ceramic tile in a tub/shower area must be fixed. If the inspection report recommends that a well and septic system be checked by health authorities prior to settlement, and a problem is discovered, the seller is responsible.

Mechanical equipment, including furnace, air conditioner, kitchen equipment, and often such things as garage door openers and fireplace dampers, must be operating. This can be an important protection for a home inspector in regard to heat exchangers on gas and oil fired furnaces, because they are so difficult to check and are such a high liability item. If there is any suspicion of a problem, the inspector should recommend that a licensed heating contractor be called in prior to settlement, noting that the furnace is included in the contract's property condition clause. If a problem exists, the seller is responsible.

It is wise to know, and to mention in your inspection report, what items are normally covered under your area's property condition clause. For example, in Florida a roof leak is covered.

It used to be that sellers could get around the property condition clause by listing a house in "as is" condition. But today such a contract is legally enforceable only if all the defects are specifically disclosed. Undisclosed defects that are discovered before settlement are likely to be the seller's responsibility.

Summary

The building inspection and real estate communities function as an interwoven network of public services. Also connected are financial and banking institutions, appraisal services, and the legal profession. Recognizing this connection as a positive development, and not viewing real estate professionals as adversaries, provides a major benefit to home inspectors.

Consider the market forces which have caused an emerging alliance between real estate and home inspection. These forces make it more important than ever for home inspectors to provide high quality professional services -- and to face the responsibilities that could make or break any small independent professional organization.

Great pressures bear on real estate agents from many different directions and commissions are not as lucrative for the individual salesperson as the general public might expect. Because there are many controls on the industry, a high level of performance is necessary for real estate professionals. Often they put time and effort into projects which don't pay off.

Consumerism has led to a greater level of complexity and Texas, the only state that has licensed home inspectors, also has a detailed process designed to control the quality of the inspection product. So far, however, the jury is still out on the question whether standards of practice have actually improved.

Inspectors who slip up are being sued. Real estate associations generally support the concept of home inspection for two reasons: because it provides a valuable customer service, and because it lessens the liability of agent and seller in a property sale.

Inspections also provide hidden benefits. Communicating these will help develop markets for you. Services you can offer include standard clauses to be used in sales contracts, being an information resource for buyers, and assuring the real estate community that you see their viewpoint in a deal. Let them know you work to identify strengths as well as defects in any home that you may be called to inspect.

Dealing with real estate agents may seem an unpaid investment in your time and effort, but it is the surest way to develop long range benefits for your business. Join the club. Talk shop over a meal. Seize every opportunity to make presentations. The object is generally to knit yourself into the business fabric of your own community.

Sample Sales Meeting Presentation

The following script is presented for your convenience. Modify and change it, using whatever applies to your situation and deleting as you see fit. The resulting personalized script will be your own to present at real estate agency staff meetings, agents' luncheons, or any time you are called on to make a presentation to a public gathering about home inspections in general.

Home Inspections: A National Trend

"Home inspections have actually been around for years, and in some parts of the country, particularly the Northeast, they have been used since the early 1960s. Home inspections have not become popular in most parts of the United States until the last few years, and are now in a growth pattern. There were good reasons years ago why home inspections were not usually done. Today the circumstances are such that home inspections are becoming an integral part of real estate transactions.

"In the 1960s, when the house they were in no longer met the family's needs, what did they do? Sometimes they remodeled, but not often. Sometimes they bought a used house, but not too often. Most of the time they bought a new house. In those days we were in the middle of the flight to the suburbs, and a new home was the way to solve your problem. New houses were of relatively good construction, and there was little need for a home inspection.

"Twenty-five years ago, the price of real estate was a fraction of what it is today, not only the initial cost of the house but the carrying costs. The actual cost of the house was low, and carrying costs were a small percentage of the family's disposable income. People did not feel a great risk in buying a house, and did not feel the need for a home inspection to protect them from mistakes.

"Twenty-five years ago, we had not entered the era of consumerism, perhaps better called the era of litigation. When somebody bought a product or service and did not get what they paid for, they shrugged their shoulders and said, "Oh well, better luck next time", and went on about their business. No one sued anyone. The liability issue had not come up in the real estate field, and real estate agents were hesitant to recommend a home inspection when it might kill the deal.

"Twenty-five years ago, the government had not entered into consumer protection. We did not have Consumer Affairs offices, and government had not started to investigate products such as aluminum wiring, urea-formaldehyde insulation, asbestos and radon. When people bought houses they did not fear dangerous products, and did not feel the need to have a professional inspection.

"Twenty-five years ago, we were still an industrial society. Then and today, blue collar workers are unlikely to pay any professional $150-200 for something they think they know all about. They think they know how to fix everything in a house, or they have a friend who does. Blue collar workers are less sophisticated buyers and therefore are not likely to ask for home inspections.

"Twenty-five years ago, houses were basic. Kitchen appliances such as refrigerators and ranges frequently were not included in the house at sale. We did not have such items as electrostatic air cleaners, humidifiers or even air conditioners. Houses did not change hands rapidly, and most prospective homeowners felt they could look at a house themselves and tell what was wrong with it; therefore, they did not need an inspection.

"Twenty-five years ago, and this is not meant to be offensive to those that were in real estate at the time, in many parts of the country real estate agents were just one or two steps above used car salesmen in professionalism. Agents were not inclined to recommend anything that might jeopardize a sale. Since real estate agents had tremendous influence on home buyers, home inspections were not recommended very often.

"Finally, in the 1960s when someone wanted a home inspection, who would they call? There were no home inspectors as such, and there was no one you could call on three days notice to go out and look at a house. At most, individual trades such as electricians, plumbers, roofers, etc. were used, or contractors were called, but they did not respond in a timely manner.

"Now let's look at the same factors as they exist today.

"Today in America, when a family lives in a house that does not meet their needs, what do they do? Sometimes they buy a new house, but not often. Many times they stay and remodel their existing house, but most commonly they buy a different used house. With 90 million-plus housing units, and only 1 or 2 million new units being built each year, it is obvious the preponderance of houses being sold are used houses. We often say that structurally sound houses tend to last forever, but integral parts wear out on a predictable basis. As houses get older, people want to know when the integral parts will wear out, so home inspections are more popular today.

"In most parts of the country, house prices have skyrocketed to 10 to 30 times what they were in the 1960s. This applies not only to initial cost but the carrying costs as well. We have 12% mortgage money, high utility costs, and taxes that continue to rise. Studies show that a family in the 1960s with a median income, buying a medium priced house, had to allocate 14% of their disposable dollar for housing. In 1989, the same family had to allocate 44%, and so people are stretched to their limit.

"Many Americans who have purchased houses have experienced appreciation in value to the point where they have almost their entire financial future tied up in their house. Because of these factors, people are inclined to view the home inspection as a small premium to pay for not making a mistake on the purchase of a house, and jeopardizing their financial future.

"Today in America, when people buy a product or service and do not get what they paid for, they sue. This is one of the major reasons that the home inspection business is becoming so popular. Today it is no longer the buyer beware, but the seller beware. Recently California passed a law stating that sellers of property, and real estate agents, are liable for defects in the houses they sell. If you don't believe that has increased the demand for home inspections in California, you should think again! This trend is sweeping the country and is a major reason for home inspections.

"Government investigation into dangerous products is raising the demand for home inspections. Some of you might remember the aluminum wiring scare of the late 1960s and early 1970s, the urea-formaldehyde insulation scare of several years ago, and lately the hot tickets are asbestos and radon. In many parts of the country, in the course of any remodeling, all asbestos has to be removed according to EPA standards. This even applies to such non-friable products as asbestos siding or roofing. In half the states, aluminum or vinyl siding cannot be put on over asbestos. The asbestos siding must be removed, adding several thousand dollars to the cost. The same restrictions apply to asbestos roofing.

"The insulation used to cover radiator pipes is almost all asbestos, and this too is incurring great costs for removal. Because of this, and concerns about radon, people now want a home inspection to make sure there is no possibility of asbestos or radon before they buy a house.

"We have moved from what is called the industrial era into the information society. Today nearly 75% of Americans are white collar workers. They are more sophisticated, but are not hands-on, and are likely to pay professionals to look at a house they know nothing about. They are becoming super consumers who want an inspection before they buy a house.

"The 76 million people born between 1947 and 1964 have now passed through the 22-year age bracket, and we are looking at a "baby bust." Because of this, many experts call for a labor shortage. I don't have to tell many of you about a labor shortage. What this means to prospective buyers is that they recognize the quality of construction in new houses and remodeling projects is less than it was 20 or 25 years ago, and they want someone to check the work.

"Houses have become more sophisticated. Appliances are included such as electrostatic air cleaners, humidifiers, air conditioners, heat pumps, and some think it takes a master's degree to run a thermostat on a heating system or turn on a microwave oven. Because of this, American buyers are less likely to think they know about a house and more likely to have inspections.

"It is often said in the next 5 years there will only be four kinds of real estate companies: one, the large national franchise such as Century 21, ERA; two, the large national companies such as Coldwell Banker or Prudential; three, the large independent broker with 20-25 offices, and then nothing down to number four, the single-office real estate company. It is no longer possible estate to run 4 or 5 real estate offices. You must run one, or twenty or more. This has substantially increased the professionalism of real estate agents, and now they recognize that in order to build their business they must have satisfied customers. Because of this they will recommend home inspections.

"Finally, in many parts of the country, and especially in your community, home inspection professionals are now available to make a home inspection part of the real estate transaction. A home inspector is equipped to check all parts of a house and to tell purchasers what they need to know to make a decision about the purchase.

"Home inspections are truly a service whose time has come. The home inspection is an all-win situation. Everyone involved in the transaction wins.

"First of all, let's look at how the buyer wins:

"The buyers win because they learn about the house before the contract is finalized, and they learn of any problems that might affect their financial investment in the house.

"If a home inspection uncovers a defect, and the buyers still want the house, they can use the inspection to renegotiate the contract or at least take the cost of repair of the problem into calculation and still go ahead with the purchase.

"The purchasers are usually present during the inspection, which allows the inspector to educate them on the care of the house. This has proven to be helpful to the purchasers and is greatly appreciated.

"The ultimate benefit to the buyer is that there are no surprises.

"There is also a benefit to the realtor. The realtor is much more likely to have a satisfied customer, who will bring referral business. On the other hand, when a major problem arises after purchase such as a bad roof, water problem, etc., even though the realtor did not know of the condition, the buyer feels the realtor is somehow responsible. The chance of referral business is small.

"If there is a problem with the house, the savvy realtor can advise renegotiating the price to reflect what has been uncovered, or persuade the buyers to should go ahead, knowing about the defect. Frequently, obvious listing errors are picked up during a home inspection, which reduces the realtor's liability. I know of a realtor at settlement whose purchaser turned to him and said "Joe, when are you going to put the storm windows on the house?", and he looked at the listing and it said storm windows. The house went to settlement $2,700 later.

"The home inspection even benefits the seller. It identifies problems that the seller may not have known about, that can be corrected to everyone's satisfaction.   It minimizes the seller's liability, particularly from hidden defects. In today's world if there is a problem, the seller is sued.

"Home inspections are becoming an integral part of the real estate transaction. It is generally accepted that before long every house sold in America will have a termite inspection, a real estate appraisal and a home inspection.

"After today's short meeting we do not expect you to recommend a home inspection on every sale, although there are agents who do that. There are circumstances where a home inspection can be used as a positive selling tool, and where it is wise to recommend it to your prospective purchaser. These include:

"If you are trying to sell a "Nervous Nellie" buyer. You may have ridden with these purchasers for two or three days, shown them a hundred houses, and at least three or four houses you thought fit their needs and on which they should have submitted a contract. But they were always afraid of taking that step.

"Many agents use the home inspection as a closing tool. When they find a house they feel meets the purchaser's needs, they suggest submitting a contract with a contingency clause based on an inspection. The point to the purchaser is that they have three to five days to mull over the contract, and have a third party explain all about the house. If you have taken any sales training courses, you will remember that any time you can change a prospect into a client, you should do it. It is our experience that when this approach is taken, and an inspection is done, the deal normally goes through.

"The second time a home inspection makes sense is when there is a known defect in the house. There might be a water problem in the basement, a noticeable structural crack in the foundation, or a bad roof. Prospective purchasers who see this problem will either not put in a contract, or put one in at such a low price the cost of repairs will be more than covered.   In these cases, a home inspector can analyze the problem, recommend a solution and most importantly, put a cost on what it takes to correct the problem. Armed with this information, the seller can adjust the price of the house or the purchaser can submit a contract with the right price to allow for correction of the problem.

"Many purchasers buy a home with the express purpose of performing substantial remodeling or renovation. With the rising cost of renovation, it is most important that a seller know what we call the X + Y equation. That is, X is the cost of the house and Y is the cost of remodeling and renovation. As part of our inspection service, we go over a house, inspect all the integral parts, sit down with the purchasers and give them budget figures within five or ten percent of what the remodeling and renovation might be.

"You might say this could be done by a contractor, who would not charge for the service. I say that if you can get a contractor to come out within 48 hours, and give a price within three weeks, I would be very surprised.

"A common situation today is purchasers who are extending themselves to their financial limit. All their ready cash is used for the down payment, and the monthly payments are pushing them to their limit. If they go ahead with the contract and do not know that the roof will need replacement in the first year, or the air conditioning compressor may go out, it could push them over the brink. The fee for the home inspection is a small premium to ensure this will not occur.

"Finally, a home inspection will sometimes help to close a sale when you have a third-party kibitzer. How many times have you almost sold a house, and on the final walk-through before purchase there is the family friend who is an architect or contractor, or the purchaser's parents who have flown into town and may be providing most of the down payment. It is easy for out-of-towners to be overwhelmed by the price of starter homes here. You have a commission riding on this, and many times cannot provide the credibility to allay fears.

"We have found that if a home inspection is scheduled, and we work as an independent expert, we can usually allay any fears and the sale will go through.

"We are finding more and more listing errors during inspection. For example, we looked at a property with a large addition listed as having a brand-new roof. In reality, it was a newly patched roof on a 15-year-old addition -- a four-ply built up roof with gravel that would cost well over $1,500 to replace. The purchasers might have taken legal recourse if they had not found out before settlement.

"We often find listings stating there is a 150-amp electrical service (calculated by adding up two 60-amp boxes and a 30-amp box), when there is actually only a 60-amp service. The cost of a new, 150-amp service is $500 or more. A home inspection will normally pick up these differences and eliminate the problem.

"You as real estate agents should not be afraid of home inspections, and should use them as a selling tool. It is important that you not sell against a home inspection. Let me tell you why.

"A couple of years ago, we were asked to look at a house. The real estate agent had told the young couple she did not think a home inspection was needed because the house was a creampuff. Fortunately, we have done inspections for friends of this couple and so they asked us to inspect this house. I did the inspection, and as I pulled up I could see from the street the roof was gone. It was an 18-year-old house with the original roof, and as you know, these roofs last 15-20 years almost to the day. I did the inspection and found little or nothing wrong except it needed a new roof, which at that time cost $1,500. The couple were sure the real estate agent knew it needed a new roof.

"I don't know whether she did or not, although every other house in the subdivision had a new roof, but I do know she lost not only the sale but a client. Professional real estate agents take the approach that if an inspection turns up a problem, they will adjust the price, renegotiate, or find another house. If you take that approach, you will not be trapped by the home inspection.

"Now we would like to discuss the appropriate role of the real estate agent during an inspection. The agent should be present. Here are our reasons:

"This is usually your first service for the purchasers after signing the contract. It shows you are following through on the sale. More important though, many times the seller will ask to go along during the inspection. If you are not there, we will say to the seller, it's up to you but we do not recommend it. Frankly, if all sellers were honest we would not have a home inspection business. If the seller is there and we say the hot water heater is 8 years old they will say no, it's only seven and a half, and we get into an argument. If we point out the deferred maintenance items, they tend to get offended. But if you the agent are there, you can say to the seller, "Come on, let's get a cup of coffee, this is the purchaser's day in the house".

"Equally critical, many times we find defects or problems in the house. As you know, most purchasers immediately react with, "Let's go talk to the seller". Again, if you aren't there we will tell the purchasers to walk out of the house, tell the seller they are waiting for the written report, and call the selling agent to handle any negotiations or discussion. If you are there, you can serve as a buffer and eliminate the problem.

"Many times a purchaser wants to negotiate based on the inspection. If you are there, you can help put it in perspective. For example, if we look at a 15 year old house with the original roof, we will say it has a normal life of 15-18 years and within one to three years it will need replacement. If the purchaser says "Hey, I want the seller to pay for that", you may be able to say "We looked at 10 houses in this subdivision. They are all 15 years old and will need a roof in the next 3-4 years. Are you sure you want to jeopardize the sale by negotiating on that?"

94

"Another example: If there is a water problem, about $500 to correct if done by a professional, and the purchaser wants the seller to correct it, you may be able to answer "You have already negotiated the price down by $4,000. This will cost $500 done by a professional, but only $100-200 if you do it yourself. Are you sure you want to risk losing this good buy on a minor point?""

"We as inspectors do not, and you don't want us to, tell a buyer how to negotiate or how to use the inspection. You are there to help buyers make that determination. We do know that fewer deals are lost and everybody is better satisfied if the agent is there at the time of inspection.

"Now, there may be times when the purchaser does not want you to go along. Many times, people have met me at the car and said, "Don't talk in front of the real estate agent". It may be you were sitting a house on Sunday, someone walked in off the street, and you wrote a contract and became both the listing and the selling agent, and you have not had time to build up their confidence. It is also a fact there are people in this world who do not trust anyone. If you were their brother or sister they would not allow you to come along on their inspection.

"It is our opinion that in such cases you should be there, introduce everybody and say, "I'll be up in the den. If you have any questions or want to talk about it when it's over, let me know."

"Now what happens at the conclusion of the inspection? Well, in 80% of cases, the buyer goes through with the contract, the contingency is removed, and that's it. A second outcome is that the buyer decides to cancel the contract on the basis of the inspection. I must tell you, you will lose sales with a contingency based on an inspection, but they will not be lost because of the inspection itself.

"Many times over the years, we have had buyers meet us and say, "I came into town on Monday, and put in a contract on this house after seeing it once. My wife flew in on Tuesday and I don't care if it's gold, she isn't going to buy this house. Can you get me out?". Well, when you have a contract contingent on an inspection by a building expert and report satisfactory to purchaser, frankly there is usually a way to get out. But in today's era of consumerism, if they don't want the house you are better off selling them another house.

"A third outcome is that negotiations are made after the inspection, with part of the repair cost to be borne by the seller in order for the deal to go through.

"One last point. We are often asked whether we guarantee the inspection. For the fee we charge, we do only a visual inspection and we cannot guarantee. There are companies that will charge an additional $200 above the inspection fee, and if they rate an item as good they will guarantee it with a $100 deductible. But if a roof is over 10 years old, they won't guarantee it. If a furnace is over 10 years old it is exempted, as is an air conditioner over 5 years old. It has a much greater psychological impact to tell a purchaser you can't guarantee a roof than to say the roof is 12 years old with a normal life of 15-18 years, and you should budget replacement in three to six years at a cost of about $1,500."

(Then ask for and answer questions.)

5. ECONOMICS OF THE BUSINESS

Where the Home Inspection Business Is Today

At the close of the 1980s, the home inspection business showed little penetration of the market around the country. Thus, the potential for expansion was phenomenal. Here is the picture, through the focus of existing conditions, which are changing daily as the market develops in our industry.

In the greater Washington DC area, home inspections are done on more than 75% of the houses purchased by the white collar middle class, the upper middle class and the rich. In blue collar areas it still remains below 25%, but it is growing. If you go from Washington DC up the coast to Boston, including Baltimore, Philadelphia, New Jersey, New York, Connecticut, and Massachusetts, the home inspection business is very popular among white collar buyers, probably well above 70% and growing rapidly. In blue collar areas it is still slow, but picking up.

If you go 200 miles southward to Norfolk and Hampton Roads, Virginia, home inspection penetration there is less than 10%, perhaps even as low as 5%. All the way down the coast to North Carolina, South Carolina, Georgia and Florida, it is between 5% and 20% depending on the area, but is moving ahead rapidly in Atlanta and in most of the larger cities in Florida.

If you move inland on the east coast to Rochester, Buffalo, Syracuse, and Pittsburgh, penetration is normally less than 15%, perhaps 25% in white collar areas. Throughout the rest of the midwest it is normally less than 10% in most areas. There are exceptions such as Indianapolis, where at one time there was a 3 to 6 week delay getting inspections, as market demand far outstripped the available inspectors.

In the cities of Minneapolis and St. Paul, there is a truth-in-housing law which requires that every used house sold must be inspected to make sure there are no code violations. This is not the same as a home inspection, but every house is at least inspected for code violations. This law does not apply in the suburbs of Minneapolis.

In 1985, a law was passed in Texas licensing home inspectors. This was supported by the Real Estate Board, and there are over 1,300 individuals licensed to do inspections in that state. The percentage of inspections is well above 70%. In general, home inspections as done in Texas may be somewhat less comprehensive than is normal throughout the country.

Before January 1987, the percentage of homes inspected on the west coast was less than 10% in most areas, though in Portland and Seattle it was closer to 20%. California has passed a law stating that selling agents and sellers of properties are liable for defects in the houses they sell, and since that time, the percentage has skyrocketed to between 80% and 90% of all houses in major metropolitan areas.

Taking the country as a whole, the percentage of inspections done today is less than 20% of the total, but climbing. It is our opinion that before long, almost every house sold in America will have not only a real estate appraisal and termite inspection, but also a home inspection.

The Market Potential Nationwide

What is the potential for the home inspection market? Here are some statistics.

According to the National Board of Realtors, the number of used houses sold each year runs between 3 and 4 million, depending on the strength of the market. In many cases more than one inspection is done on an individual house. This means the market potential is 3-1/2 to 5 million houses.

At least 1 million new housing units are built each year, depending on the strength of the market, and of this number approximately half are single family dwellings. In many cases more than one inspection is done, as there is a close-in as well as the final inspection. This means the market is somewhere between 1 and 1-1/2 million for new homes.

In 1987 there were approximately 100,000 condominium projects in America, with 10 million condominium units. The CPA policy group that audits the books of condominiums has recommended strongly that an audit of the condominium association's books on an annual basis should include an inspection of the physical facilities, a review of the maintenance procedures, and an analysis of the reserve capital to ensure they are adequate. This means that every condominium project needs to have an inspection at least once a year -- a minimum of 100,000 inspections, and these are not the normal $200 inspections, either.

There are nearly 5 million commercial, institutional and industrial buildings in this country. They change ownership, on an average, every 7 years. Commercial inspections are becoming more and more popular, particularly with the introduction of subjects such as asbestos removal. It is likely that in the next few years, all commercial buildings that are transferred will be inspected, giving a market of 500,000 or more inspections a year. These are more complicated and higher priced than home inspections.

There are at least a million lawsuits each year regarding housing: between builder and new home purchaser, between owner and remodeling contractor, and so on. In every such lawsuit, at least two inspections are required, one for the plaintiff and one for the defense. If the case goes to trial, expert testimony is needed on each side of the case. This means there are a minimum of 2 million inspections required each year, plus the possibility of expert testimony appearances.

As the home inspection business grows, inspectors are gaining a reputation as objective third-party experts. When owners of property, homes or buildings or condominiums, wish to undertake projects or have a number of bids they cannot evaluate, they call for an inspection on a specific condition. Partial inspections are quite common in the home inspection field, on such things as roofs, structural problems, and water problems. The inspectors inspect the condition, analyze the problem and recommend a solution, in some cases including how much it will cost to correct and writing specifications for the repairs. We estimate there are a million of these potential inspections per year.

There are in this country over 500,000 paid corporate relocations per year, and 1.5 million unpaid. This is one of the major sources of home inspections. People coming in from out of town are more likely to ask for a home inspection.

As interest rates fluctuate, millions of homes are being refinanced. More and more, refinancing applications include a home inspection. It is likely the number of re-financings will exceed 500,000 each year which is another opportunity for home inspections. Home equity loans are also likely to include a home inspection.

Government agencies, national, state and local, are recognizing the high cost of maintaining inspectors on payroll. If inspections can be done by private outside inspectors and paid for directly by the user, it is more efficient and cost effective. One county in Virginia instructs builders to get their own new-home footing, close in and final inspections. They are told they must have a registered engineer, architect, or certified BOCA inspector, who will do all three of these inspections, and be paid by the builder.

In one area of Ohio, a home inspector does all the inspections for remodeling permits and splits the fee with the local jurisdictions. There could well be a market of over 500,000 to 1,000,000 inspections in this category.

Companies are developing hazardous materials specialties and becoming expert on inspecting for asbestos and radon in houses and buildings. The asbestos removal market alone is forecast at $2 billion, and 65% of commercial buildings built before 1979 had some asbestos. This is going to represent an enormous market, not only for the initial inspection to discover the material but the write-up of specifications and the monitoring of removal. Some experts believe that soon nearly half the United States will have a radon inspection on houses sold, which can mean 1.5 million radon inspections per year.

One large insurance company in the United States will no longer write insurance policies on any commercial building that has ever had asbestos. They are concerned that great liability still exists if the removal process was done improperly. We believe that most insurance companies will insist on this, not only for commercial buildings but for private residences. We foresee the day when an insurance company will require an inspection before writing a homeowner policy, to discover asbestos, urea-formaldehyde insulation, aluminum wiring, whether the roof leaks, the pipes are likely to freeze, or the electrical service is unsafe, and to verify the presence of smoke detectors. This could represent an enormous market.

There is presently a total market for more than 10 million inspections per year, and this number may be vastly understated. This means a potential for 500 inspections per year per individual, in an industry of 20,000 or more inspectors in the United States, by 1992. As of January 1990, there were fewer than 6,000 individuals doing inspections on a full time basis. This means the market has a potential of 5 or 6 times growth.

Structure and Policy Demands

The economics of your business as a home inspector will vary greatly depending on how you approach the business. Running a very small operation from your home differs in some significant ways from a larger operation involving office space, full-time staff, hired inspectors and other considerations. First let's examine the common areas that apply regardless of the specific size of the operation.

Unlike remodeling contractors or any of the construction service businesses, home inspectors have a high degree of operating control over their work. You can run your policies and make the individual jobs fit your procedure, which need not be changed because one client is overly demanding.

While the time demands are intense, depending on your level of energy you have the advantage of being able to close the door at the end of the day and pursue other interests. You have flexible-inflexible hours. This means you can set the hours to be "on" or "off", but during the hours you are "on" you are inspecting houses, writing or dictating reports, or doing other tasks directly relating to the business 100% of the time. When you're not "off", you're "on".

The general specialty of your business is home inspections. This involves several major types of jobs. They include:

√ Structural/visual inspections of pre-owned houses

√ Punchlist inspections of new houses, code inspections during construction, code inspections for resale of existing homes

√ Inspecting multi-family units for owners' associations, inspecting condominium units for individual purchasers

√ Inspecting commercial buildings, warehouses, etc.

√ Partial inspections for various purposes such as court testimony, and appearing in court to testify in lawsuits

Your contract is by simple fee. Keeping the fee structure clear and simple is to the advantage of all concerned. This schedule should be set as an indicator of the complexity (time is your major cost) of the job.

The scope of your work includes electrical, mechanical, utilities, plumbing, structure, weathertightness, grading and grounds, and by specializing you can also include certain more technical areas such as asbestos and radon testing or other environmental areas. These latter categories require a different approach and will make it impossible to deliver reports complete within 48 hours of inspection, because often lab work will be required to test for asbestos and radon.

Your major risk is liability due to oversight and errors. In addition you risk time over-runs, personal injury on-site, and travel time and road hazards.

You may occasionally encounter problems relating to scheduling when people don't show up, the utilities aren't turned on, or weather conditions prevent your complete inspection of HVAC equipment. Unlike the remodeling contractor who has any number of possible delays due to unforeseen problems, your job problems are truly minimal.

The ownership and cost of equipment is a fraction of the investment required of even the small shop or remodeler. Their interest on capital, withholding and insurance costs for employees, will more than pay for your entire inventory of equipment. In a small office, the tools required to do the actual inspection -- a vehicle, office equipment, and dictation equipment -- are all you'll need. By using on-site forms you can eliminate almost all the clerical time needed for a written report.

If you run one or more inspectors in addition to yourself, you can operate with minimum tax and insurance costs by setting up your inspectors as subcontractors. This means the inspector receives a percentage of the fee after the company has collected, and you don't require specific office hours when the inspector must be present. You can use a simple cash method of bookkeeping, either doing it yourself or by subscribing to a service.

The key to making this business comfortable and profitable is to understand what you will face in the day-to-day operations and to establish policies to meet those circumstances. The establishment of those policies gives you easy control over the business and frees you to do the best inspections possible.

A Small Office Operation

One of the major benefits of the home inspection business is financial control. By comparison, the remodeling contractor has to go through a complicated estimating process with variables that are not always apparent, such as weather conditions or the expectations of the client. Getting a grip on that situation is possible, but requires continuing alertness and an ongoing question mark needling in the back of the contractor's mind. Direct job costs, mark-ups, and correct estimating procedures are a constant pressure.

In your home inspection business, however, direct job cost is almost always clearly predictable, because it is the percentage of the total paid to the inspector. This varies between 25 and 50 percent depending on the policy you set. Thus, your gross profit will range between 50 and 75 percent. The most common job cost or inspector fee runs between 35 and 50 percent of total gross revenues.

Let's assume you are operating on a 60 percent gross profit, which means that you are paying your inspectors 40 percent. Out of that gross profit comes your overhead costs. They are likely to be within the range that follows:

General Management -- five to 10 percent. If you are running a company with three or four inspectors and doing $400,000 worth of business, the general management figure should be $20,000 to $40,000. This is what the general manager should earn for running the business, training people, and so on.

Advertising and Marketing -- five to 10 percent. Marketing pays off more than is generally believed in the home inspection business. It includes advertising, direct mail, and personal calls on real estate companies. Many companies pay the inspectors to do it.

Rent -- one to two percent is a common figure. While office space doesn't have to be in the high rent district, there are times when you need a presentable office space to meet with business associates. If you are running your operation as a business, you should budget this amount.

Automobile -- two to four percent. If the owner or manager wants to include automobile as an overhead item, the cost will vary depending on volume. Total cost of the automobile, including $250 to lease or depreciate the automobile, $100 per month for insurance, and $150-$200 for fuel, oil, and maintenance is reasonable. This adds up to $500 or $600 per month, $6000 to $7200 per year. If the company were doing $400,000 per year in inspections, the automobile would be about two percent. At $200,000 gross, it comes closer to four percent.

General Insurance -- usually one or two percent. This is for general business and liability insurance. It doesn't include workers compensation on inspectors because they are generally subcontractors, and it doesn't include errors and omissions insurance, which is a separate item.

Secretarial -- four to six percent. This includes the cost of staff to set up appointments, type reports, and handle office clerical work. If all your inspection reports are typed or generated by office staff, that cost could run from $15 to $25 per report and run your secretarial costs as high as ten or twelve percent of gross.

Office equipment and supplies -- two to four percent. This varies depending on whether or not you use a computer, do narrative reports, or buy printed forms for on-site reports and job management.

Credit cards -- one percent. Credit card income normally runs about 33 percent of the gross, and has a service charge of about three percent. Thus the total credit card cost is one percent of gross.

Bad debts -- less than one percent. While the home inspection is usually paid for on the spot with cash, check, or credit card, there are instances when it must be billed. Bad debts from this category normally run under one percent.

Education and training -- one to two percent. A budget of one to two percent is recommended for keeping abreast of developments and polishing your skills, probably more if the company is small.

Accounting and legal (excepting claims) -- one percent. Because the methods and issues are fairly clear, the cost is low.

Errors and omissions insurance, including legal fees -- six percent of gross plus $2500 deductible. This is a typical cost, but it can vary widely. Depending on insurance policies available at the time you need coverage, it will pay to shop around. In many cases, self-insurance is the answer in the form of an overhead item of five percent of gross revenues, held out as a contingency against the day when a claim is made against your company or against you personally.

Because of the high likelihood of a claim, it is very important that you be prepared to meet that unpleasant reality head-on when the day arrives.

The Independent Inspector

If you are running your business out of a home office and doing all your own inspections, the picture is a little different. Operating expenses are low, and relatively more can go into your own pocket. Here is the picture:

Management -- five to 10 percent. While this number is correct, in your one-person operation it is something you pay yourself, and thus you pocket the money.

Advertising and Marketing -- While you need telephone service and Yellow Page ads as well as some direct mail and brochures, many times the company is small enough that your main marketing effort costs you primarily time and shoe leather. You're dealing with your personal reputation, and marketing doesn't necessarily have to represent an out-of-pocket expense.

Secretarial -- four to six percent. While this is a normal figure, you can handle most if not all of this personally if you have the time. Using a telephone answering service and pre-purchased on-site reports, you can cut the paperwork to a bare minimum.

Rent -- In a very small operation, from your den at home or as a sideline out of the office you use for another business, there will be little if any cash outlay for this item. The tax ramifications should be understood and planned in advance, however, as the IRS hears bells when a home office deduction is filed with your 1040.

Automobile -- While the automobile for a small company would run five to six percent, it could also be considered a job cost since most inspectors pay for their own automobile.

Errors and omissions -- varies widely. This may be hardly noticeable when you're operating your own small service, but it could also be very high if you had a large claim placed against you early in the game. The answer is to prepare yourself not to make errors and omissions, and to operate as carefully as possible. Horror stories abound of inspectors who were ruined by undertaking a job they didn't fully understand and then not doing it right. Be conservative. The risk you take is your own. Budgeting 5% for legal costs and claims is prudent. One inspector in California sets aside 10% of each fee into what he calls his "extortion fund."

Income Potentials

From the perspective of a relatively large home inspection company with over 20 years' field exposure, HomeTech can report that overhead runs between 30 and 40 percent of gross. This reduces to a net profit of between 10 and 20 percent depending on what the field inspector is paid. For very small companies, overhead normally runs no more than 20-25 percent.

If someone did 300 inspections in a year at $200 per inspection, this generates a gross income of $60,000. If this inspector were also the owner in a small operation, subtracting overhead of 25 percent would leave a total income available of $45,000.

If inspections increased to 500 per year, using the same relative numbers, subtract 25 percent of $100,000 to achieve a total available income of $75,000. The one-man operation could add a second or third inspector and continue to run on a relatively modest basis. You can easily see that the numbers top out well above $100,000 for this kind of operation.

HomeTech recently heard from a home inspector in the northeast whose remodeling company had gone bankrupt the year before. He said, "The home inspection business is great! I do 30 inspections and gross $6500 a month, working 24 hours a week. Since I work out of my house, my overhead is only my car, telephone and marketing expense. I couldn't be happier. This certainly beats the 80 hours a week I worked in remodeling -- at less money!"

Fee Structure

Fees vary substantially throughout the country, from as low as under $100 for code compliance inspections in the upper midwest, to $200 or better for established businesses in areas where the inspection industry is going full tilt. HomeTech believes that a typical bare minimum anywhere in the country should be $125 to $150. In Minneapolis and St. Paul the truth-in-housing mandated inspections run under $100, but are not the same sort of service a typical inspection would entail. State officials in Texas have suggested that a good quality home inspection is worth about $200, and the fee is necessary for inspectors to stay in business over the long run, although some are working for less today.

There are a number of ways to price inspections, some more complex than others. One that is easily understood by inspectors and clients alike is based on the selling price of the property. For example, $150 minimum fee on houses up to $150,000 and $1 per thousand in cost above that to a total cost of $250,000. Above that, the cost increases 50¢ per $1,000. The schedule looks like this:

Contract Price of House	Cost of Inspection
Up to $150,000	$150
$175,000	$175
$225,000	$225
$250,000	$250
$300,000	$275
$450,000	$350

This is derived by subtracting $250,000 from the house value and rounding off the figure to the nearest $1,000. Then take 1/2 times the number of thousands and add to $250.

$450,000 - $250,000 = $200,000

200 x 1/2 = 100

$250 (break point) + $100 = $350, derived fee for a $450,000 house

The fee trend is upward, and as liability increases it is reasonable to see a minimum fee in the $175 to $185 range. Unless you are just starting out and are seeking to expand the market, this is a good fee to shoot for. Otherwise, don't expect to stay in business long at $150 or less, but use the lower fee temporarily while you generate initial repeat customers.

Another fee structure is based on the square footage of the house. This can be confusing, but it runs along these lines:

Size of House	Cost of Inspection
Up to 1,000 square feet	$125
1,000 to 2,000 SF	$175
Over 2,000 SF	$225

When the cost-per-square-foot fee structure is used, misunderstanding is the frequent result. People ask whether garage or basement, decks, and porches are included. If the fee is not clearly understood when the inspection is scheduled, you are risking difficulty when time comes to collect.

The inspection fee should be kept simple so that everyone involved knows and can easily remember it. The underlying principle is to offer no surprises, but create a predictable situation.

When you use a sliding scale, it is better to quote a fee on the telephone and make it a specific dollar amount so the real estate agent understands, the buyer who is writing the check understands, and the seller is aware that the service being rendered has a significant dollar value.

As prices of houses rise into the $200,000 range and more, the fee quote is based on the effort involved to make the inspection. Some areas of the country have houses valued at $1 million or more. In this circumstance, the time and effort involved should not call for a fee in excess of about $500.

Fee Variations

In multi-unit dwellings, a different approach is called for. A house with a separate English basement, or a two-unit apartment, might be charged the basic fee plus a reasonable add-on, such as $50. Apartment houses are normally negotiated, but a good thumbnail guide would be $400 to $500 for a 20-unit apartment house selling for $500,000 if you inspected just the common areas and four or five units. If the request is for every individual unit to be inspected, add at least $30 per apartment to the fee.

Partial inspections. These are limited to a single item such as a roof, water problem or structural problem. A reasonable fee would be $25-$35 less than your standard fee. While this may appear high to newcomers, the costs in time and travel, the written report, and after-inspection consultation come into play. You can't make money by charging less than $125 to $150.

New houses are normally done under the same fee as resale homes. If a customer wants more than one inspection for a new home, as is often the case, the first is done for full fee and a discount applied to the second. Even though an initial close-in inspection takes less time than a final punch-list inspection, the full fee is collected up front to cover your initial time and energy.

Condominium inspections are charged at the same price as a regular home inspection. The fee involves checking the individual unit and the overall project. This can be quite complex, including structure, electric, plumbing, heating, and the financial status of the condominium association as reported by the real estate agent. The inspector is trying to discover whether there are conditions that could cause a rise in the condominium fee in the near future.

Re-checks. Some companies don't charge for re-checks done as part of the initial inspection. For example, if an inspector goes to a house where electrical service and water are not turned on, he will ask the purchaser to get these turned on and then will return to finish the inspection. Other companies charge $75 for this re-check because it takes away time the inspector could have used for a different inspection. HomeTech doesn't charge for re-checks unless a specific trip at a specific time is required. The reason is, without the re-check the inspection has not been completed, and so the full fee hasn't been earned.

Cancellations. It is impossible to charge for cancellations in the home inspection field. If a real estate agent on Monday feels sure a contract will be signed on Tuesday morning and calls for a 1 PM inspection, only to learn early Tuesday that the deal has fallen through, the agent has no choice but to call and cancel.

You can't fairly bill the agent for having had the foresight to anticipate a signed offer and arranging for your company's service. To do so would be to lose the agent as a referral, and you still wouldn't be likely to collect. Fortunately, appointments come in at all hours and so your canceled appointment may well be filled by another. As much as 10-20% of your appointments will typically turn into cancellations.

Mishaps. If the key doesn't work, or nobody shows up for an appointment, HomeTech's approach is to reschedule the inspection and charge nothing for the inconvenience. Call it good will or public relations, the fact is we are in a service occupation and completion of the service is the only realistic basis for charging the fee.

Legal testimony. A typical fee for legal testimony is to charge the original inspection plus $60 to $75 for preparing for the courtroom appearance, and a minimum of $400 per half day, even when you're sitting and waiting to be called. Many inspectors around the country charge $100 an hour for expert testimony. It is important to get the money up front, since if the ruling goes against your client or the client doesn't like your testimony, you will not be able to collect after the fact. Ask for a retainer, then collect a check before you take the witness stand.

No matter what fee you set and the client agrees to, the judge in a case may rule that a maximum witness fee will be paid. If this should ever happen, you would have no choice but to comply.

Collections

Collect your fee at the point of sale if at all possible. This reduces the accounting burden and eliminates a lot of bad debt. Sometimes you can't do this and will have to live with the results. Let's hope the worst result you experience is a delay of 30 to 45 days in making the collection.

During the 1970s, HomeTech billed clients and carried a large accounts receivable in the company books. Then one morning in 1979 a quick review led to a startling discovery: over $60,000 in unpaid accounts receivable! We learned at that point to make all collections on the spot. It is easy to do by setting a clear policy and training staff to include the following wording when making an appointment:

"The fee is payable at the time of inspection. Do you wish to pay by cash, check, or credit card?"

It's easy to set up an account at your local bank to allow you to accept VISA, MasterCard and American Express. You don't need a machine and there is no up-front cost. All you have to do is get the full name of the customer, credit card number and expiration date. You call the charge in for approval, deposit the signed slip at your bank, and they credit your account. Once a month the bank deducts two to three percent (depending on the charge card's rate) and that's it.

Real estate attorneys in some states handle the inspection and pay for it out of escrow at settlement. If you accept work on this basis, you may have trouble collecting if something happens to prevent the deal going through. You've done your work and delivered the report, and you are legally due the bill rendered. But there is a chance you will never collect.

This should not deter you from developing a good relationship with several real estate attorneys, because if you are a member of the team you will get repeat business and in all likelihood, the attorney will remember you deserve to be paid and help collect the money for you.

Paying Your Inspectors

It is much easier for a company to hire inspectors as subcontractors. To preserve subcontractor status, you must draw a line between the services required of an employee and the fees paid for a service rendered. You can't give subcontractors the paid vacation benefits, on-the-job training, medical insurance coverage, or require regular daily attendance at your place of business that you routinely do with employees.

In addition to easier bookkeeping because of the various fringes that don't apply to subcontractors, there is a tax savings to the company of 7.65% Social Security, .7% Federal unemployment, and state unemployment that may range from 2 to 4 percent. You don't pay retirement benefits to subcontractors and you don't usually pay a regular vacation.

At this writing there is some question whether there is any way the IRS will be willing to view home inspectors as subcontractors. The IRS' Circular E, Employer's Tax Guide, says the following:

"Anyone who performs services is an employee if you, as an employer, can control what will be done and how it will be done. This is so even when you give the employee freedom of action. What matters is that you have the legal right to control the method and result of the services."

Under the circumstances, the best recommendation is for you to get legal advice on your own company situation.

The inspector should be paid only after the job is collected. HomeTech pays its inspectors on Friday for money collected through the previous Monday noon. Even though most of the work is paid on-site, the occasional bill that must be waited for should be reflected in the inspector's check as well. Pay after collection, and make subcontractors aware that this is the policy which will always be followed.

How much to pay the home inspector is a question that companies can experiment with. Some pay beginning inspectors just 25% of the amount collected. These companies generally have dissatisfied inspectors who consider working for such a small piece of the gross to be time-in-training.

The inspectors are aware that they are working for themselves anyway, and so they are looking for a personal following, perhaps feeling out other companies for a better share, or waiting for the chance to go into competition. There is little incentive, and no feeling of loyalty or any sense of being vested in the parent company. Companies who are stingy with their key people pay a high cost in personnel turnover.

As the inspector's fee increases to the range of 35-40%, equity can begin to be felt. The inspector should be on contract, and the contract should spell out potential improvements in the financial share paid to the inspector. Usually this progress is defined by the number of inspections done for the parent company.

One method which works well is the policy HomeTech finds successful. The sliding scale looks like this:

Number of Inspections Done	Percentage of Gross
1 - 100	40%
101 - 349	45% when inspector is requested
350+ (award a territory)	40% as inspector, +6% for marketing and requests

The inspector must share in the company's liability risk. The HomeTech policy is to take 5% off the top, before percentage shares are figured, that goes into a claims fund. In addition, inspectors pay 50% of each individual claim up to $1,000 (including legal fees).

Summary

The home inspection business nationwide has a potential for five-fold growth or more in the next few years. This is because the market potential has barely been recognized, much less developed fully. Market penetration today is about 20 percent.

If home inspections were fully realized today, there would be about 500 inspections per inspector per year potential nationwide. If an inspector nets $100 per inspection, the expected national average, full employment, is $50,000 annual income. For entrepreneurs who establish their own businesses, there is a greater profit possible.

A one-person operation taking a fair share of the national market could realize an income of $60,000 to $75,000 annually. Running an office with additional inspectors, better than $100,000 annual income could be achieved.

Economics of the business show a variable overhead of 30 to 40 percent for large companies, resulting in a net profit of 10-20 percent or higher. In a small operation, overhead can be as low as 20 percent but will more likely be 25 percent, giving a before-tax income of 65-75 percent of gross.

A holdback for errors and omissions is the sleeper. While specifics are impossible to predict accurately, a contingency of about six percent is a reasonable cost projection.

Fee structure should be kept as simple as possible. Charges normally are based on the cost of the house or on square footage. It is less confusing to use a cost-based fee, because everyone can remember it easily. The best policy is not to surprise your client with an unexpected fee. While the difference of a few dollars is not likely to make or break your company, it could chase customers away if they are offended.

You will do several types of inspections over the course of a year, ranging from partial inspections for a specific problem to whole condominium projects. By planning ahead, you can get a good control over how to base your charges for the different kinds of buildings you will inspect.

As a matter of policy, you should be in business to please the customer. Be willing to accommodate and bend your rules slightly if necessary to achieve this goal.

When hiring inspectors, it is preferable to use a subcontractor relationship if possible, for easier bookkeeping and slightly lower overhead. The important factor is to protect the inspectors with a progressive share of the gross to develop loyalty and make them feel good about the company.

Remember, your inspectors could become your competition. To avoid this, use enlightened management practices and reassure your people that they are valued members of your company team. Pay inspectors based on money collected, not on jobs which have been billed.

6. MARKETING, ADVERTISING, PUBLIC RELATIONS

Bootstrap Image Making

Let's face it: the small scale home inspection business is a personal business. People learn about you as an individual and think of you or an associate when they are considering a home inspection. Forming the image for this kind of operation is shoe leather-intensive. You may not have a huge and sophisticated marketing/advertising budget, but you will have significant time invested getting your service before your buying public.

Most of this will involve personal appearances wherever and whenever opportunity permits, from real estate sales meetings during business hours to burning the midnight oil writing articles for your local newspaper.

You're literally marketing at all hours of the day. When you do an inspection you have a small audience with the potential for you to make four sales: to the purchaser, the seller, the seller's agent and the purchaser's agent. Since much business comes by personal referral, don't ever underestimate the importance of these contacts. The agents are not direct consumers but brokers who will recommend you to the consumer. Buyers and sellers are your consuming public.

Whether you are in a rural or urban area there are techniques you will need in order to be effective. There are media where advertising can be counted on to generate business, and certain kinds of promotional materials and ways of creating them that will repay your time and effort. You must find out which are likely to benefit your situation. Some elementary market research is necessary.

This chapter gives you the tools you need to analyze your marketplace, and shows you how to address your image making efforts in an effective strategy for success.

It should be clear by now that real estate agents are the direct source of much of your business, and it is appropriate to concentrate a lot of your marketing energy on these real estate professionals. If you choose to look at real estate agents as your adversary, then your marketing must be directly to consumers. This is very expensive, and not very productive.

Two home inspectors in Canada went into business together, and after a year and a half they were doing one or two inspections per week. They used a 15-page detailed, technical report and they used only direct-to-consumer advertising, on which they had spent almost $15,000. After consulting with HomeTech, they changed to narrative reports of 3 to 5 pages and began marketing to the real estate community through seminars and sales meetings. Within a year they were doing 7 to 10 inspections per week, and their business has continued to grow.

A financial institution went into the home inspection business, giving a free inspection to homeowners who bought a mortgage. The firm's advertising agency persuaded them to introduce their service through radio, TV and newspaper advertising, but the results were almost zero. Real estate agents would not recommend a home inspector they had never seen, even if the inspection was free. The moral of this story is that real estate agents should be your allies, and treated as such.

A real estate agent who sells $2 million per year will be involved in about 20 transactions per year. If you can gain the confidence of 50 such agents who will recommend you as one of three inspectors offered to clients, you have a chance at a thousand inspections per year. If you get only your one-third share, you have the opportunity for 334 inspections, which is not bad for a year. There are many home inspectors who spend a lot of their time at four or five real estate offices, and gain most of their business just through that effort.

Basic Tools

Industry-wide experience has shown that you need certain basic tools to create the image of any small business. In your case, these tools will include business cards and forms, and a brochure that can be stocked in various places such as the Chamber of Commerce, real estate sales offices, and perhaps local motels. You should have a name for your business, and an advertisement in the Yellow Pages. The cards and brochures should be systematically if not professionally designed.

Additionally, you will need: the ability to write a news release or article for your local paper; a prepared presentation with slides that you can customize for special interest groups; and a sample of one or more inspection reports in an attractive and easily-understood format that will show potential clients what they can expect once they engage your services. This should be the same report format you use in your regular business.

For anyone familiar with writing a home inspection report, the news release or feature article will be no problem. Simple guidelines are all you'll need, and they are included later in this chapter. The enthusiasm and thought process required to create additional articles, or a potential series, should come naturally to any home inspector. And the image you can create by making this relatively small and enjoyable time investment will be invaluable to your business over the long run.

The idea is not to advertise your business specifically, but to create a climate and market for home inspections generally. This is market development of the most elementary kind, but it is sure to pay off. As the expert with the by-line, you'll earn a significant share of the market that evolves.

A presentation using video tape or slides is easy for any home inspector to deliver. Even if you're not a seasoned public speaker, by using visual props that interest you and your audience you'll soon have everyone involved in the subject. After a few times, you'll probably look forward to the stimulation of making a public presentation that starts everyone thinking and asking questions.

If your community has a historic preservation group, your slides can be customized to focus on historical details of houses and how they need to be treated in the preservation effort. These may be anything from architectural features of cornice, column, and miscellaneous trim to roofing techniques and materials, foundations, garden walks and walls, porch structures and roof shapes.

You can use many of the same slides for a presentation to the real estate board at their monthly luncheon, or the local bar association, where attorneys who handle real estate (and order inspections) will appreciate your viewpoint.

You may use the same set of slides or props on different audiences, but your message should be customized to the main interests of each group. Questions from the audience may be quite different after you've offered essentially the same presentation to groups with divergent perspectives.

Although you may not consider it a marketing tool, the report you deliver is tangible evidence of your service. It is the product that people take away with them. You may not be able to give a good example other than by direct experience, taking a client or group through an inspection and pointing out the things you observe.

But you can show examples of your reports to groups so they will know what to expect from you. Then they can grasp your meaning when you describe your generalist approach, or the technique of creating perspective by making negative observations on the "uptake", when you're describing positive features as well. You are in a service business, but the tangible report is what people will remember you by, and it is a marketing tool.

A number of inspection companies use "gimmicks" to keep their name visible, such as pens, calendars, memo pads, flashlights, etc., to give to real estate agents. Some companies have labels with their name to paste on furnaces and tags to go on main water cutoffs that they leave as they go through a house. Out of sight is out of mind in the real estate business, and even a small object with your name on it can help.

Market Research

If you're going to develop a sound marketing strategy, it must be based on realistic market research. The potential market size must be determined for your area, and then market penetration can be assessed. Following that, you can consider market share, which is the portion of the current market that goes to each company doing business in the area.

In a small area, or a larger area where you are just starting out, a day spent making visits to various offices will pay off for you. You'll become better known and begin to develop relationships (always carry a couple of brochures and business cards to hand out on these visits). And you'll learn some important things about the potential for your success, while developing strategies to reach your marketing goals.

Potential Market Size

The potential market in your area will be the total of several different items. One is total home sales for the year, a number you can get from the local real estate board (a non-governmental group). Many real estate agents will tell you frankly how many homes are on the market in multi-list at any particular time.

Visit with a real estate agent and, after a few minutes to break the ice, simply explain that you are researching the market for home inspections in the area and need some data. Their confidential listing book reports the total sales with breakdowns by area. This book comes out twice monthly throughout the year. By referring to a late December book you can determine the total sales for the year in the multiple listing area.

Breakdowns for "commercial and industrial" and "lots and acreage" are also given. To determine the number of residential sales, subtract the sum of the two sub-categories from the total. Allowing for agencies not fully disclosing their sales to the Multiple Listing Service, and for sales by non-member offices, you can add about 10%.

Condominium association inspections are required according to some association by-laws on a yearly basis. You can determine the number of condo associations by talking to the local Chamber of Commerce. Then you'll need to make a reasonable guess as to how many potential multi-unit inspections exist. This kind of inspection takes a certain amount of salesmanship. You'll need to judge how well you feel you can do with this market segment.

A visit to your local Building Inspector's office can result in useful information. Ask how many permits for new home construction were issued in the previous year, and how many permits for residential renovation/remodeling. Ask whether the office does its own inspecting, or whether builders are accountable for a certain portion of the inspection process.

If there is a chance that builders certify the inspections were done and the structures comply with codes, you may be in a position to do up to four partial inspections for each house that is built. Ask how you can become accepted by the Building Inspector's office as a qualified private inspector. It may simply be a matter of using certain approved forms available for the asking, of registering with the office, or other formalities. Then your market size can expand significantly.

As a public agency, the Building Inspector's office will reveal certain information about permits. In most cases the name and address of the property owner will be included whether or not a contractor actually took out the permit. If the owner has an out-of-town address, there is a potential for you to do progress reports and inspections on the structure whether or not the Building Inspector does the code inspections during construction.

A letter to the owners, with a copy of your brochure and a simple explanation that you do more than minimal code inspection, has a high probability of getting business for your company. The information you gained has given you a highly qualified prospect, something the professional marketers are always keen to develop. Even if the names and addresses are local, a letter wouldn't do any harm and might gain you some business.

Despite the overlap between houses sold and permits issued, or creation of new housing and sales of housing, the information you have gained with a few visits has given you the data you need to determine the potential market size for inspections in your area. Add up the subtotals to find out how large the universe is. This doesn't yet give you a realistic market size, but it is an important first step.

Before you can project the likely size of your business, you need to know market penetration, or the use of home inspections in general in your area. You can best find that out by talking to an officer of the local Board of Realtors. The figure you get won't be exact, but it will be an educated opinion.

Remember when talking to any real estate professional that the National Association of Realtors is really the godfather of the home inspection industry. This doesn't mean you should be timid about asking for a few minutes to discuss home inspections with a local real estate professional. It does mean that real estate people tend to be educated about home inspections and to understand that inspections can be excellent selling tools for them.

If you want to backstop the real estate professional's opinion regarding the use of inspections, make another stop at the Chamber of Commerce and talk to the executive director there. You may not get any firm answers, but you can get a valuable insight. Based on these two visits, you should have a percentage figure with which you feel satisfied.

Actual Market Size

Multiply that percentage times the potential market size you learned earlier to determine the approximate market in your area. Your visits around town have begun to get your name and your presence known. In this effort, your market research also functions as personal image making, or shoe-leather marketing.

If your research shows that there are 1600 total potential inspections in the geographic area you feel you can serve, but only a 25% market penetration, then you have discovered two important things. First, there are only about 400 inspections waiting to be done in your area under present conditions. This is the actual market size. Second, you have to get at least 50% of the 400 inspections in order to make a full time living as a home inspector. This means you will need at least a 50% market share from the outset.

You will need to devote significant energy to market development, probably by doing a newspaper column and possibly a local cable channel program on house problems, which are not difficult or expensive, but are time-consuming activities. So while making your rounds on this personal market research trip, better include the local newspaper and the cable television office as well.

Market Share

Now that you know what is required to make a going business of home inspections in your community, go to the Yellow Pages and look under the heading "Building Inspection Service". If there are several firms offering to do inspections, call each one and ask about their service. You can probably get the information you want by asking about the kind of inspections offered, the cost, and how many inspections per year the company does in this area. If you ask "blind" and can get the information without pushing for it, as if you were shopping for an inspection service, so much the better.

Listen closely to the replies given, and make notes. Let your competitors try to sell you their services, and see whether you can think of a better way to respond to the very questions you are asking. If your competitors are frank and honest, and embellish the facts just a little, you can probably believe they are doing half to two-thirds as many inspections as they claim.

Call each number listed in the directory and ask whether inspections are a full-time business or part-time. It is very likely that you're either the first in the area with serious intentions about an inspection business, or that you're well equipped to provide a strong alternative to the established businesses. If the competition isn't serious about the business it is highly likely they haven't done any marketing homework and you already have a better concept of the potential.

If there is an existing market for 400 inspections and the established companies seem to be doing about 175, do the math:

175 divided by 400 = 43%

This means the existing competition is only taking a 43% market share. You already have a good potential as the more professional company by making a more effective presentation and adopting a sound marketing strategy!

Setting Goals and Adopting Strategy

In all likelihood, your marketing goal will be to establish your company by earning at least 200 inspections per year. Since this is a large percentage of the market we have "discovered" in this example, you need an aggressive strategy to convert perhaps 20% of the competition's business to your side and usurp the additional business you know exists.

Chances are that contractors are doing home inspections "on the side" with no real system or knowledge of the business, although they do have a first-rate understanding of houses and their problems. And a portion of the remaining business is probably going to an inspection company from a nearby area. Inspectors from this company must drive long distances to serve this business, and it is barely practical or profitable for them to do so.

This situation is challenging. You need to call on all your resources: time, energy, professional service, marketing skills. Your strategy will include regular personal calls on the largest real estate companies, public presentations to local civic clubs and real estate board meetings, regular "coffee stops" at the building inspector's office, the newspaper office, perhaps the local radio and television cable stations. You'll need to buy some paid advertising space and possibly follow up on personal rounds with a direct mail campaign.

You'll have to create about half of your business and count on conversions for the other half. While home inspection is an "easy-in, easy-out" business which is not capital-intensive, you know from your market analysis what to expect regarding profitability. It may take a few months. You may have to keep two incomes coming in for the next year or so.

Conversion vs. Creation

There are two ways to get your market share of the business, by conversion, or winning over the existing business of a competing company, and by creation. Conversion is easier.

First, the established home inspection company will have a set way of doing things, and can't be quite as flexible as a newcomer with an open mind and the determination to "work harder and do better".

Second, every established company has a percentage of slightly dissatisfied to nearly resentful customers who are itching for a change of pace. These are almost always real estate salespeople who feel stung because an inspector didn't put quite the right "spin" on a report, and may have caused problems or even been instrumental in the loss of a sale. Regardless of whether this feeling is realistic, they are prime prospects for conversion to your service.

Third, any business going to the established company is already aware of the home inspection service, thus it is a developed market. Fourth, if the existing companies don't know the size of the market, there are gaps in their marketing program that you can turn to your advantage.

It won't be quite so easy if you're unknown, a stranger in town. Under these conditions, you should become seasoned by working for another inspection company for a year or two. Going into an undeveloped area as a new, independent stranger will make your job tougher, but it can be done. In fact, home inspectors today in most parts of the country are starting from ground zero, having to sell not only themselves but the whole idea of home inspections.

The best place to begin a new home inspection business is where many companies are already operating, because the conversion of existing business will be relatively easy. Conversion won't take nearly so long as creating the market in an undeveloped area. And if the real estate market is fairly active, you will find plenty of sales agents willing to try a new home inspection company.

Creating or expanding the market beyond its current size is more demanding, and slower. Since you have no control over the size of the potential market, the process requires that you increase market penetration. This will involve as many marketing and promotional tools as you can manage, from public service and personal appearances to paid advertising in local media.

In most parts of the United States, a new home inspection company must educate both consumers and real estate agents on the benefits of a home inspection. Do not underestimate the challenge of this task. Even though circumstances, like the California law that declares selling agents "legally liable for hidden defects in the houses they sell," have given tremendous energy to the home inspection business, there are still many agents who "don't want anything killing my deal!"

123

Your challenge is to take every possible opportunity to explain the positive benefits of a home inspection. Some opportunities include the following:

✓ Offer to give presentations to real estate offices on commonly found products or conditions that may require negotiating the cost of removal or mitigation, such as asbestos, radon, FRT plywood, aluminum wiring, basement water problems, and bad roofs. The goal is to establish yourself and your company as an expert information source.

✓ Offer to every service and garden club in your area a talk on subjects such as:

"The Six Most Common Defects Found When Buying a House"

"Home Inspections: a National Trend"

"How to Buy and Fix Up An Old House"

✓ Become part of the training process for new real estate agents. Offer to give a 3-hour presentation on "The Anatomy of a House" or "How to Inspect a House for Listing and Sale." Use slides to teach the difference between asphalt and slate shingles, what a 30-amp electrical service looks like, how galvanized pipes rust, etc.

✓ Sponsor seminars to the public on housing topics. For years, HomeTech sponsored a 12-hour, 4-session seminar called "How to Buy and Fix Up An Old House." This established the company's reputation as housing experts, and about one-third of the participants were real estate agents, many of whom frequently called on HomeTech for house inspections.

✓ Take a booth at a home show. You are quite likely to get enough interest just from other exhibitors such as real estate companies, mortgage brokers, remodelers and home builders to make it worthwhile. Try to make your rental of the booth contingent on your giving a seminar at the show on almost any subject related to housing; this is another good way to establish yourself as an expert.

Relocation Market

People moving in from out of town, especially from the more urban areas, are likely to request a home inspection. There are several ways for you to tap this business. Most national real estate companies, franchises, and larger local companies have relocation specialists. These are good people for you to know.

There are also national relocation organizations who buy homes from large corporations. The corporations purchase the homes from their employees who have been transferred, as part of their benefit package. Another benefit that a number of corporations offer is to pay for a home inspection when an employee is transferred. One home inspection firm got all the business from a corporation that transfers 400 people in and out of the city every year.

Relocation inspections often must be billed rather than paid on the spot, and radon and asbestos inspections as well as estimated cost of repairs may need to be included.

Condominium Associations

Inspection work for condominium associations can be an enormous opportunity. Most condominiums require an annual inspection to review maintenance procedures, condition of physical plant, and necessary reserves for capital expenditures. The very worst phrase in a condo association's vocabulary is "special assessment." Your expert inspection provides the information needed for appropriate monthly fees that will cover large expenditures such as a new roof, exterior repainting, etc.

Your marketing plan should identify local condominium associations and management organizations, and keep in touch with them through personal contact and direct mail.

Publicity and Public Relations

Publicity and public relations are the "soft sell" component of your marketing strategy. You don't often have the chance to make a direct appeal for business, but your efforts can be more cost effective and, in the long run, more likely to bring in business per dollar's worth of effort expended.

The appeal is to the general public, and the message is how home inspections benefit buyers and sellers alike. Intermediate subjects include how to spot trouble before major damage or health hazards develop, how to maintain and prepare a house for seasonal changes, chasing out the "ghosts" that go "clank in the night", and various items relating to plumbing, electrical, and heating systems and equipment. These can be the basis for a newspaper column, a radio talk show such as "Ask the House Doctor", or a cable television series.

Shaping Client Expectations

An excellent subject in a public relations campaign might be "what to expect when your home inspector calls". One of the thorniest problems in home inspection is how to meet client expectations. And the worst situation a home inspector can encounter is a dissatisfied client whose expectations were not met. To such a customer, minor oversights can be magnified into major errors.

The problems you solve by making your service well understood by the general public may be uncountable, but your efforts will benefit your business and the public at large. Most likely they will help prevent lawsuits by giving consumers full information going into an inspection.

This process can be broken down into three elements:

√ Tell them what they are going to get

√ Give them what you told them you would

√ Tell them what you gave them

Public Relations vs. Publicity

All public appearances are educational public relations, whether mini-seminars for sales meetings, lunchtime stand-up routines, radio or television guest slots. You're interacting with other people, and you are learning as well as teaching. Materials you develop for news and feature columns can be either public relations or publicity. Subject matter features about things other than your specific business are public relations. You are serving the public good by preparing these.

News releases describing your business activities by name and in the third person, announcements for the business column or feature page, are publicity. Consider reporting whenever you attend a training seminar or home inspector's convention, when someone new joins your firm, or when you receive a special award or other recognition.

When you plan to give a seminar or make a public speech, send in an advance notice. Always follow up with a two-page release on the content of your talk.

Since you control what you will say, you can have it on the business editor's desk the same day you make the presentation. This is the kind of timely coverage any newspaper editor appreciates, and whether or not the item appears in the paper this time, you will begin to establish predictability and reliability. As a consequence you may be called for quotation occasionally, or you may find the editor receptive to your ideas for a weekly column.

Both public relations and publicity take careful thought and effort. They take time, but they don't cost a lot. Publicity and PR establish your presence in the marketplace, project your image as a civic-minded business person, and lend authority to your service and opinions.

This market development benefits the entire industry by making home inspections more visible, and will contribute to expanding the market in general. When you make the pie bigger your business is sure to increase, even if your market share doesn't change.

Using media exposure effectively. Media available for no cost other than your time include local newspapers, shoppers, magazines, and newsletters: the print media. Participation in these pages will require a written release and possibly a photo package or line drawings. The electronic media, including radio and television, require your presence at a studio at the very least.

Feature articles. Consider inviting a reporter/photographer team along on a home inspection. This is a popular angle for the newspaper because the reporter is the "eyes and ears" of the general public and editors may be confident of their colleague's impartiality and style. For you this means taking a little longer on the inspection because you are still working for the client, not the reporter, and must put the client's priorities first.

After giving your report to the client you may need to go back, with the client's or owner's permission, and set up photo shots. Take the time to do this, because the publicity you'll gain is better than any advertising space you could buy. The key to this promotion is your client's willingness to accept some invasion of privacy.

If you don't feel comfortable asking this of a paying client, try offering a free inspection to anyone on the newspaper staff to provide a background for the story. It will cost you two or three hours, the newspaper will get a human interest feature, and the staff member will get a "free" inspection. You'll make friends of the reporter, photographer, news editor, and the homeowner. Your payback will be far greater than the return from any paid advertising bought with the fee from one inspection.

Don't change your emphasis or technique for the newspaper inspection, but show everything that you would show to a regular client, and give the normal field report and follow-up report. Make the experience totally authentic, and your stature and credibility will be greatly enhanced.

Interviews. To get a newspaper interview, you will need some advance preparation. Try using your market research as a "news hook" for the interview. Think of a catchy way to phrase the report of your market findings. Something like, "A new local survey indicates home inspections are saving buyers thousands of dollars and major headaches their first year after purchase." Another likely subject: "Ten most prevalent house problems in this area revealed by inspection survey."

Whatever your research reveals, generalize the findings and report the results clearly and succinctly. Don't reveal your methodology or the numbers of respondents to your questionnaire. While it may not be scientifically accurate, the fact that a survey was taken will be local news. Tell the editor about your results and offer to provide a list of concerns you can discuss for publication. Make up a list of simple statements and be prepared to answer the editor's related questions.

As a professional home inspector you will be recognized as a quotable authority, especially if your research relates to local conditions and says something that hasn't been reported in quite the form you can offer. You'll be assured coverage if you can bring up issues of money, safety, comfort, or how local conditions differ from the general perception ["Chimney fires, not faulty kerosene heaters, cause major damage locally"].

News releases. If you're in a small metropolitan area served by a single local newspaper, or two small competing papers, you can prepare news releases on almost any non-commercial subject and stand a good chance of having them printed. The interview subjects above could be cast as news releases, or you might write announcements on the business page as described earlier. Some likely subjects include:

Feature/Column Items

Local contractor discovers crawl spaces have silver linings. (How your part-time inspection business has taken off, and what your customers have to say about your service.)

Things the building inspector didn't mention. (Contrast code inspections by public officials with paid home inspections, emphasizing home maintenance and systems familiarization for owners.)

Simple fire prevention for houses with aluminum wiring. (Describe measures the homeowner can take to avoid fire hazards.)

Cause and solution if your paint is peeling. (Discussion of different causes for paint failure and suggestions how to prevent recurrence when the house is repainted.)

Chimney leaks, and how to stop them. (Sealers, flashing, repointing, etc.)

Stop pouring money through the cracks. (Weatherization techniques that apply locally.)

Don't shun the sun. (Simple techniques to take advantage of solar gain and increase comfort in your home.)

Basic basement fitness. (Controlling water drainage with minor regrading.)

Rating your roof. (Basic roof inspection and troubleshooting.)

How home inspections strike a balance. (Describe how inspections identify major strengths and weaknesses to present a balanced description of the house, avoiding minor nit-picking.)

Publicity Announcements

Inspectors add employee. (Announcement of personnel changes in your company.)

Home inspector completes training (attends seminar, etc.)

Local business adds division (when you expand an existing business to offer inspections, or new services.)

Company initiates new television series (radio show, etc.)

Business donates services to a worthy cause (give a free home inspection to someone contributing to public radio, public television channel, etc.)

Print Media Campaign

Project description. You can go on endlessly thinking of subjects for business announcements and news releases. Make a list of ten subjects for a news feature to add to the items presented above. Your list of 20 subjects, which you can research easily from your course and field experience, can form the basis for a continuing newspaper column. You'll need two written columns and the list of additional subjects when you present a package to your local newspaper editor for approval.

Project goal. Your goal is to create additional market for your services by producing a local public service column.

Strategy. Begin by talking over the idea with your newspaper editor. If you reach a meeting of the minds, ask specific questions and follow the instructions carefully. The questions include:

How many words long should each article be?

Should we use artwork when possible? If so, what kind is most appropriate?

How often would you use this column? How many articles should be given to you in advance, and how many should you have on hand? This is another way of asking what the copy deadline is, stated in terms that the editor will find comfortable.

If the editor says a 500-word count is appropriate, don't turn in 700-word articles. This makes it more difficult for the copy editor to fit the material to the reserved or "budgeted" space, and will probably result in your last 200 words being lost. The easiest cut for a newspaper editor to make is from the end of the article, so you can't write toward a conclusion but must state your most important information as soon as possible.

The media pitch letter. In a larger area where the newspaper is run on a daily schedule, and business is more formal, you may begin communicating with the news media by sending out pitch letters. In this case, your writing sample is out in front, and the letter must open the door for you. It must be clear and dramatic. Here's a sample:

Dear _____ :

Can you imagine a worst case scenario for a young couple buying their first home? The financing has just made mincemeat out of the family savings and required major restructuring of the monthly budget. When finally they move in, the following discoveries come in rapid succession:

The powder room toilet flushes with hot water, but the sink doesn't have hot water at all . . . The basement floods every time it rains . . . Heat and air conditioning don't get to the two end bedrooms . . . Seven wall receptacles don't supply electricity . . . And the dishwasher stops in mid-cycle.

These and other typical problems can be avoided. With as much as two out of five days on the job required just to make house payments, young homeowners can't stand the additional burden of hassling with contractors and making major repairs.

By using the services of a professional home inspector before final settlement, home buyers can take the worry and guesswork out of the house purchase.

Our inspections, done before the purchase contract is finalized, examine all the major house systems to give a true reading of conditions and likely cost of necessary repairs. Our service saves consumers hundreds, even thousands of dollars, and avoids nasty surprises that can haunt homeowners for months or years. I believe readers in our community would be very interested to learn more about the benefits of home inspections, and have enclosed a brochure to describe our > > (company name) < < services.

To help inform consumers, I'd like to offer you some non-commercial "house help" articles. Enclosed is a list of (quantity) article ideas that I'd like to discuss at your convenience. I'll call you shortly to discuss the concept, but if you have comments or questions in the meantime, please give me a call at (your phone number) .

Sincerely,

Preparing the package. Don't forget to include your list of article ideas, your company brochure, and your business card. If you have clippings of articles on home inspections, include one or two photocopies.

Then be sure to **follow up.** Mark your calendar, and call within one week to set up an appointment. In most cases, this approach will get the editor's attention and result in a good interview for you.

Electronic Media

With the expansion of cable television, there is a huge demand for programming with a local or community service angle. Your expertise should make you an ideal host or guest on a show. The same is true for local radio programming.

Radio is easier. Check local broadcasters to see which shows have a call-in talk format. Then approach the producer or station manager for a discussion. Present your credentials, and offer to appear as a guest to answer listeners' questions on house problems. There is a very good chance your offer will be quickly accepted.

You could end up a local media personality, carrying the image of the home inspection industry to new readers, viewers, and listeners every week. Of course, this effort will bring great rewards to your company as well, as to the public, through consumer education about houses and the benefits of home inspections.

Paid Advertising Where it Counts

Paid advertising includes business cards and brochures, direct-mail advertising materials, and paid space ads. Paid space can be display advertising which uses a design created by an ad agency or the publication where the ad is to appear, or it can be lines of copy in a classified section or telephone book, or it can be time bought on radio or television.

You may need some of all the above categories, but you will need brochures and Yellow Page advertising to demonstrate that you are actually an established business.

Yellow Pages

This opportunity comes along only once a year, so you will have to plan in advance. If you're considering starting your own business, call the telephone company business office for instructions on how to contact the Yellow Pages advertising representatives.

These sections of the telephone directory are often handled by independent companies who sweep their sales force through the countryside like a military campaign. You may not be able to get a direct answer on the cost of various size ads until the salesperson calls on you, but at least register with the company so that you will be contacted.

Usually, if you have a business telephone listing you will qualify for one single line insertion under the Yellow Pages category of your selection at no additional charge. You can increase your exposure for about $16 per month with a small ad, on up to a relatively large cost for a display ad. The fact that you are billed monthly for a publication that isn't distributed monthly may seem unappetizing, especially if you add up the total charge for the small space over a year. But there's no way around this system, so your choices are limited.

Look under the classification "Building Inspection Service" to see how many competitors are operating in your market and what sort of advertising they are buying. Then you can do the same, or buy a slightly larger ad if you can make it distinctive. The most important thing in a display advertisement is the headline. The second most important item is the telephone number. Everything else pales by comparison, so don't try to make a sales pitch in the directory ad. The goal of this investment is to get people to call you. If an ad is large enough to do that, it's large enough.

Other directories. You may consider listing your business in the city directory, a hardback volume that many businesses use to compile lists of consumers for one reason or another. It also lists telephone numbers numerically and location by address. The city directory is often carried over by businesses from year to year, and you may wish to purchase space in this only once every other year.

There are trade directories you might also consider. One is aimed at the relocation industry, called the National/International Relocation Directory. Write to them at 1901 Avenue of the Stars, Suite 1774, Los Angeles, CA 90067.

Listing in these publications won't bring you immediate business, but over the long run they might open up some channels that will benefit you.

Company Brochures

The company brochure is your most important marketing tool. It should be interesting, attractive, professionally designed, and available in every real estate office in your trade area. Begin to compile the information you will use in your brochure by researching the competition's offerings. See how they present their services, then decide what you can offer that distinguishes your business from the rest.

Some areas you might consider for brochure coverage include:

Type of service offered: general inspections, construction inspections, technical measurement, radon/asbestos/environmental hazard inspections, accompanied inspections. If you specialize in taking your client personally through the house to point out items needing routine maintenance and how to operate equipment, maybe you want to point out that your inspections include free "informative review of house systems for the new owner."

Speed of service: If you use an on-site report, which is attractive to real estate agents for negotiation and closing the deal, make prominent mention of the fact. If your report is followed up with an extra free telephone conference at the buyer's convenience (during office hours), this increases the perceived quality of your service. If you offer a detailed written report mailed or available for pick-up within three days of the inspection, mention that as well.

Extra handouts and services: This is good for a photograph in the brochure. If you have a home maintenance book with space for recording data and keeping warranty information, try to cover this without too much description, but make it attractive. Free consultation is a good thing to feature as well.

Professional qualifications and membership: If you're a member of the Chamber of Commerce, Better Business Bureau, Board of Realtors, etc., list these items in a clear block of space. If you hold professional training certificates, include this information. The point is to establish you as a solid business person with impeccable qualifications.

Company motto or identification line: If you have a good motto, people will remember you and you will earn calls as a result. A few examples you may use or improve upon:

"Soothing the home seeker's troubled brow"
"Protecting the house hunter from disaster"
"Showing new home buyers the ropes"
"The home buyer's shock protector"
"Professional inspections to prevent purchasers' aftershock."

Personal experience and company data: If you have over 500 home inspections in the area but are a one-man office, skip the personnel and concentrate on the number and type of inspections. If you have letters of commendation from clients, excerpt from one or two and include them. If you write a weekly column, do a television or radio show, or make appearances describing the benefits of home inspections in any manner you can describe in a few words, include this as well.

Headlines: Remember, your brochure sells from its appearance. Your text should not be too long. List the most important benefits you can deliver. When you've gotten 10 or 15 items, go back over the list and boil down each one to no more than three words. From this you should begin to see good potential headlines. You'll only need four to six heads in a three-fold brochure (8-1/2" by 11" paper folded twice), so be selective in the words you choose.

Logo and illustrations: You should have a company logo. It may be professionally designed for you, or selected from a catalog or government publication. Use it on stationery, business cards, report forms, and your brochure. Major creative effort is often invisible to the naked eye, so don't make the mistake of thinking your logo has to be complex and subtle in order to be good.

For your brochure, line art is often good enough to convey the idea of houses in a neighborhood setting. One good approach is to show a section of a house on the cover and a labelled view through the structure on the inside, with points of interest identified. This progressive style of illustration gives your benefit message a dramatic punch.

Personal photos can be useful if you have two or three inspectors. Be sure to include each inspector, with names, if you include a photo of yourself. You're selling a company, not an individual. You might also include a photo of yourself with a clipboard and a client looking where you're pointing on a house.

Once you have the artwork and copy, it's a good idea to employ a graphic artist to put the package together for you. Don't just hire any artist you come across, but review their work. Be confident that they do good brochure and logo art, and that they understand your business and the flavor of your message. Since you're both in service professions, you might even offer the artist a home inspection as payment for the work.

The time spent developing this sales tool may be a five-year investment. It will more than repay you before you need to redesign the brochure. As services change and people come and go, you can update it using the same format. Graphic design does go stale after a time, but good design looks fresh much longer than second-rate work.

Other Image Makers

An important part of a professional image is your on-the-job presence. Dress professionally, on a par with the real estate agent and the buyer. Drive a clean, neat, late-model vehicle. If you carry tools and equipment, keep them clean and neatly packaged. Don't get to a job and go rummaging for a widget in the trunk of the car. These may seem to be minor considerations, but they are important in the overall image that you are creating. Remember, in addition to performing a public service, you are forever marketing!

You can buy electronic media time and space in the consumer press, but the response cost is quite high. If you budget $100 a month or so to keep a small advertisement in the local newspaper's real estate section, this will eventually bring in business. However, the public has to see an ad at least half a dozen times before it will register if the consumer isn't actively seeking your service. After another three or four exposures, you will begin to get accurate name recognition, and perhaps the individuals looking for a home will make a note to call you for an inspection when they buy.

If a real estate agent mentions you along with two or three others, and your ads have made an impression, then you stand a good chance of getting the call. Money spent on a great brochure will also swing business your way when several companies are presented by an impartial agent.

For paid advertising space to be effective, you've got to be in for the long haul. If you intend to place ads for just a month to "test" your response rate, save your money. You'll be disappointed. Local advertising, like everything else in business, must be consistent and persistent to succeed.

Summary

Marketing your wares is a full time undertaking, and you must be alert to how you are coming across. It isn't necessary to have large dollar investments in public relations and marketing, but it's vital to have planning, goals, and a strategy to help you reach the goals. You need personal presence and other basic skills to be your own effective marketer, but it is a personal business, so your winning personality is the most important factor.

Your marketing efforts should focus on your becoming the expert information source in your community for anything connected with housing. The more credentials you can acquire, and reputation you can establish to support this image, the better. Teaching courses, presenting seminars, giving speeches, being active in local civic and professional associations, are all good steps to take to build this position.

While you may believe you are in a strictly service business, the public views your handouts as your product. Your inspection reports and other literature you provide have got to be memorable, satisfying, and useful. The handout package is your single most important marketing tool, combined with your personal contacts and presentations to real estate agents, offices and sales managers.

Market research is a good way to introduce yourself to important groups and businesses that will work with you if you have material they can use. These include civic groups who want to hear what you have to say about common house defects, the newspaper and other media in your area, the Chamber of Commerce, real estate groups and others.

As a writer of inspection reports, you are naturally inclined to be a good writer of news feature articles. This is both market development and public service. It will generate goodwill for you to be a regular newspaper columnist. To pursue this you must do your homework before you approach the busy news editor. Thinking about your business and the services you provide will be good discipline. Listing such things will form the basis for a clearly thought-out editorial proposal.

There are protocols you should know and follow, including the pitch letter and simple preparation of press releases. These include format of the page and structure of the article. News writing style is similar to report writing, but you must rearrange your sequence of presenting the facts to meet the practical demands of the news format.

In working with electronic media, you have the advantage of being in a profession that is easily televised. You could work up an inspection show with a minimal amount of equipment and personnel, but a lighting crew is very important. Your presence on television is a significant factor and will affect the delivery of your message. Radio is easier than television, and you could begin a series of guest spots on a talk show with call-ins to the "house detective". All these efforts are indirect generators of sales for your service, and in time they will repay your efforts.

The company image is most often carried out with a logo and a brochure. These are best developed by professional designers, but your input will make the difference between an exercise and a lively sales presentation. Make your lists, present your work to the artist, and get the design professional involved with your excitement about home inspection. Then you're almost certainly assured of a successful promotional piece.

Your personal demeanor and appearance are important aspects of your company image and your marketing strategy. You can try buying ad space in general interest publications if you have staying power and a budget you can allocate for at least six months, but don't expect results overnight. It takes time and repeated exposure to get the attention of the general reader.

7. ADMINISTRATIVE PROCESSES: MAKING YOUR BUSINESS SURVIVE TO THRIVE

Importance of a Good Business Plan

Picture yourself as a building contractor setting out on a job that will take your major attention for the next six months and result in about $25,000 profit to your company. Would you do it without a detailed set of plans? Without specifications? Without careful take-offs and calculations? Without a well-written contract? And if you tried to eliminate the planning work, how long can you imagine your company would survive?

Chances are you wouldn't even know you were "on the skids" until the registered letters from suppliers and their attorneys started arriving. There would be no way to "play catch up" because your structural design would be simply intuitive and based on how well you could remember from day to day which commitments you had made and how you expect to meet future commitments.

Eventually, you would find that you had never been really in business at all. Perhaps you were engaged in an interesting but ultimately sobering exercise in self-delusion. Bankruptcy would be the likely result, and the self-styled independent business person could face public ridicule as well as financial disaster. And all for a lack of planning!

Now picture yourself as you would like to be one year from now. Make a mental image of your office, its surroundings, your co-workers or employees, the kind of equipment you will have. Picture the car you'll drive, see yourself on a job handling the remarks of a satisfied client. Look into the future and see a healthy bank balance with your name on it. Hold these images in your mind with eyes closed for one or two minutes.

Ben Vitcov, a former building contractor who turned to home inspections in 1980, is now the president of Property Inspection Service, Inc., in the San Francisco Bay area. His service performs more than 8,000 inspections per year and ranks in the top 5 percent of home inspection companies in the nation.

In Vitcov's words: "I've been in business since 1957 and have known many people who went into business and failed. However, I have never known any business to fail that was run with the benefit of a business plan. Obviously, the preparation of a business plan is not a guarantee of success, but it will dramatically improve your odds."

People neglect to write a business plan because they don't understand the process, or they mistakenly think they know what they need to do, or they simply do not understand what business they are really in. Making a business plan, especially the first one, isn't remarkably easy. But the effort is always worthwhile. You'll learn things about your operation that you perhaps dimly imagined before. You'll have the information in a written form that you can analyze and compare.

It took Mr. Vitcov 100 hours to write his first business plan, but he considered the time well spent. His yearly plan now takes 20 to 30 hours.

What is a Business Plan?

A business plan is a detailed, written statement of the personal and monetary reasons for establishment of a business, expectations of its accomplishments, the methods of its operation, and the procedures to determine its success.

In writing your business plan you are not imagining. When you create the mental and mathematical construct in written form, you establish the pattern for your own behavior and the operation of your company. Not only will your expectations be formalized, but your mental goals will affect your daily attitudes. You will find it easier to maintain a winning attitude if you keep your goals in mind, and your business plan will help you to do this.

In the first place, you must decide to Make That Plan. The work involves a lot of self-analysis and serious thinking about your own operation and the realities of your marketplace, competition, overhead, marketing strategy, finances and other components. You could spend the better part of two weeks making up your first plan. And then every year, you will need to update and modify your strategy to conform to the goals you have set for your business.

In the final analysis, this kind of thinking about how to run your business is the framework on which you hang all your daily operations. If you're going to be in the home inspection business, you need to know many things, more than simply the best methodology for making an accurate inspection.

Set a deadline for yourself and make every effort to meet the deadline. While there are various approaches to writing a business plan, just as there are various forms for a financial statement, the intended effect is always similar. You anticipate variables in the operation of the business and create strategies to turn every situation into an opportunity.

Part of the plan is to show anticipated results of certain actions. Another part is to demonstrate to a banker, investor, or partner that you have a sound business approach and a good chance to achieve success.

You can think of your business plan as a presentation of your business proposition. A scientist would write a grant proposal, where you write a business plan. The difference is, your plan is going to show a bottom line profitability, and be presented in terms that people in different areas of the commercial community will accept and understand. As a home inspector, you're well equipped to create a document that communicates both concept and strategy.

Description of the Business

The first part of your plan should describe the business, its name, location, purpose, and owners/operators.

You may feel this formality unnecessary; after all, you have these details in your head. Maybe you think of your new business as Home Inspection Associates one day and Home Inspectors Association the next. When it's time to file corporation papers, open checking accounts, apply for loans, buy advertising, or tell your secretary how to answer the telephone, you must be consistent. This applies to other details of your business, as well. The business plan is your point of reference about every aspect of your organization.

Name, address, and location. Use the exact title of your company. If it is incorporated, don't forget the Inc. at the end. The company name should not be your own personal name unless you will never add any more inspectors. If your company is John Jones Inspections, and you recruit Jim Smith, he can never be John Jones. But if the name is something generic like Professional Home Inspections, Jime Smith can comfortably be an associate of that company.

Give the street address, city, post office box number, if applicable, and telephone number. Indicate the general location of your business and specify areas you plan to serve. This will not limit your operation, of course, if you later wish to expand the area served. You will soon learn that all business plans are made to be revised.

Description of business. Describe the exact nature of your business. State any limitations of your home inspection service; for example, you may not wish to inspect apartment buildings or condominiums. If you plan to include commercial inspections, state the types you will do, such as retail stores, small manufacturing plants, office buildings. Also list any limitations as to total square footage, number of stories or height of buildings, or age of buildings.

Include the types of clients you will seek and contacts you may use to locate them, such as real estate brokers, trust attorneys, loan officers of financial institutions.

Describe briefly the financial basis of your operation -- cash only, accounts receivable, credit cards -- and when you expect payment by the customer. Indicate your policy of using written contracts and any exceptions you will make.

Operation of business. State the type of organization you plan to set up, for example, a one-person operation with an office in your home; a partnership with another home inspector or with someone in a different line of business; a corporation; a subsidiary or department of another company you own.

List the officers and managers of your company. Indicate if you will hire a staff and the types of positions you will include, or if you will utilize subcontractors as home inspectors, secretarial and bookkeeping services, or telephone answering agencies.

Include a statement about any aspect of your business that you feel should be a part of your company's policy on your first day of operation.

The purpose of the business. This section will contain your reasons for operating a home inspection service. Your first reason undoubtedly will be to make money, but you probably have other purposes. These could include performing a valuable service for home buyers; utilizing your talents to increase your self-satisfaction and community prestige; or educating the public about home maintenance and safety.

Business Philosophy and Objectives

The manner in which a company deals with its customers, employees, stockholders, subcontractors, suppliers, and the general public reflects the business philosophy of its owner or owners. It becomes an integral part of the policies and procedures of the company. There are numerous factors involved in the success or failure of a company, but an important intangible asset of any successful enterprise is its philosophy and how it is applied to everyday operations.

Goodwill is a term used to describe the results of an appropriate business philosophy. Goodwill has no monetary value and cannot be carried on the balance sheet as an asset. However, if a company is sold, goodwill often acquires a dollar value and can be included as an asset by the new owners.

A statement of philosophy is a necessary part of your business plan. It directs you toward your goals and becomes a vital part of business management.

OBJECTIVES OF YOUR BUSINESS

In accordance with your business philosophy, you should list the objectives of your company. You may wish to include specific figures for sales and profits for given periods, or you may prefer to make general statements. These could include:

Customers:
- Quality of inspections
- Timeliness of inspection reports
- Good communications
- Honesty and integrity

Employees:
- Working conditions
- Pay scales and benefits
- Career opportunities

Stockholders:
- Good management
- Adequate return on investment

Subcontractors:
- Fair and honest practices
- Prompt payment
- Team atmosphere

Suppliers of goods and services:
- Prompt payment of invoices
- Honesty and courtesy

General public:
- Community involvement
- Encouragement of employees' participation in community events
- Honesty in advertising
- Education in home maintenance and safety

Managing the Business

Now you must decide how management is going to work and who will make it work. If you are a one-person operation, the answer may seem fairly simple. But do you have the technical, management, and clerical skills to operate a business by yourself? Will members of your family help you? Or will you require the professional talents of an accountant, secretarial service, advertising agency, or answering service?

If there are others involved, is there sufficient expertise among you to manage the business? What responsibilities will be assigned to each person? Will you require outside help in some areas?

THE MANAGEMENT TEAM

Be specific about who will do what, and make these decisions a part of the business plan. Who will be responsible for answering the telephone and scheduling appointments; who will keep the books; who will handle marketing and advertising; who will type the reports and verify they are written in accordance with the agreements and standards?

Write down every function of the business and make sure all bases are covered before you start. You don't want to lose customers because the telephone is not answered. And you certainly don't want to deal with an IRS auditor because no one is keeping the books.

If your business is incorporated, you may be the president. But who will be vice president, secretary, and treasurer? If you are a partnership, who will be responsible for management of the operation? If you divide the duties, be specific about each person's job.

IN THE FIELD

As a one-person operation, you will handle all inspections. State this in your business plan. If you expect a volume increase or if plans include more than one inspector, indicate whether you will hire field staff or utilize subcontractors. If you do not wish to make that decision at start-up, note that it must be made prior to expansion.

As a partnership or corporation, you may have more than one home inspector as owners. If there is a preference as to types of inspections or if one person is more qualified in a particular type of inspection, these should be included in your business plan. Also indicate who will be in charge of the field work.

Perhaps you wish to enter the home inspection business, but do not feel you are presently qualified in every aspect of an inspection. You may anticipate contracting with a specialist in a given area, such as electronic security systems, roofs, or swimming pools. Indicate this in your business plan and, if possible, line up the people you will need and include their names.

IN THE OFFICE

If you have an outside office, your management decisions include choosing its location, the type and size of office, its personnel, and its operations.

The choice of a site for a home inspection business will not require as much research as many other businesses, since the actual service takes place in the field. However, there are several considerations:

- The cost and other aspects of the lease or rental agreement.

- Is the office on a bus route and does it have parking facilities for employees and visitors?

- Does it offer a clean and safe working environment?

- Are there sufficient amenities to impress a potential client in accordance with your business standards?

You should also include in the business plan a list of furnishings, equipment and supplies required for the office, any decorating or renovations that will be your responsibility, and, most importantly, your office staff requirements.

Include designation of the person who will manage the office. You may do this yourself, but consider the time you will be away from the office and the necessity for a staff member to act for you.

Market Analysis

Before you can write your business goals, interest other stockholders in your corporation, or approach a bank for a loan, you must perform a market analysis.

WHAT IS A MARKET ANALYSIS?

You learned how to conduct market research in Chapter 6. A market analysis consists of:

The data you gathered: demographics of your area; the number and type of potential customers; what is usual and customary in local home inspections and fees; and information about the operations, sales volume, profits, and types of inspections of your competition.

Compilation of these data into workable figures to indicate the present and future status of the home inspection market, in sales volume and gross profits.

Determination of the market share available to your company, based on total available sales volume and the type and number of potential customers.

Methods you will use to gain your share of the market and to create new markets.

RESEARCHING THE COMPETITION

A market analysis will include as much data about each competitor as you can obtain. You may wish to complete a form for each company, including:

- Sales and profits
- Percentage of market
- Number of years in business
- Strengths and weaknesses
- Customer satisfaction
- Quality of work
- Fees and operational methods
- Inspection specialties

MARKETING STRATEGY

Your marketing strategy should be included as the summary of your analysis. Approach to your market share will usually take two basic directions:

√ Converting your competition's customers into your customers

√ Creating a new market through:
Personal involvement in the community
Promotional materials and advertising

Education through newspaper and magazine articles, public speaking,
    radio and television programs
Displays at home and trade shows and other local events

The best public relations expert you ever employ may be yourself. Remember, every contact you make could mean a new customer. Keep this in mind as you develop your marketing strategy.

Business Goals and Strategies

By now, you have firm ideas about the operation of your business and how it will fit into the marketplace. It is time to write down your goals in terms of measurable performance, including dollars and cents.

Goals serve several purposes:

√ They motivate you to attain a successful business operation.

√ They help you to determine the strategies required.

√ They keep you on track; you know where you're going and how to get there.

√ They furnish a basis for the financial requirements of your business.

WHAT TO INCLUDE IN YOUR GOALS

You will use the business plan, as developed this far, to set up your goals:

• Your philosophy and objectives indicate what you want to make of your business.

• Your management decisions furnish the basis for its operation.

• Your market analysis shows you how your business fits into the local economy.

You are probably most interested in projected sales volume and expected profit figures. Before you make money, however, you must spend it.

Your first consideration should be an estimate of monthly expenses, how many months of cash reserve will be required for each item, and total start-up funds required for each expense. See Sample Form "Start-Up Funding" on page 267 for a worksheet to help you determine these figures.

This plus your market projections will help you to work up your volume and profit goals. Be as realistic as possible. Your goals should challenge you but not be beyond your means to accomplish.

Goals may include any aspect of the business you wish to track for months or years. They need not be the kind of goals that can be assigned volume or dollar figures.

Net profit and return on investment are important, particularly if you establish your business as a corporation and wish to attract outside investors. These goals also are vital to a financial institution's decision to lend you money. However, in a service business like home inspections, unless you are personally known and respected, there is a high likelihood that the banks won't be your most enthusiastic supporters from the very beginning.

Production goals, or the number of actual home inspections you anticipate performing, should be included. If you plan to do all the inspections at first, a goal could reflect future business expansion. This goal could include the estimated date you expect to add other inspectors.

If you plan to operate your business from your home, but look forward to opening an office in a business location later, this goal should be listed. It may include estimated date, projected overhead expenses, and personnel requirements.

You may to incorporate other activities into your business such as public speaking or participation in seminars to educate consumers about home inspections, maintenance, and safety. You may wish to qualify as an environmental hazards inspector. Expansion of your operation is limited only by your imagination and should be an integral part of your goals.

How to Write Your Goals

Goals should be established for specific periods. The Small Business Administration recommends three-month, six-month, one-year, and five-year goals.

Each goal should include:

Priorities. Certain actions will be necessary to start up and operate the business. Other actions may be potentially beneficial but not crucial to the business, especially during the early days of your business.

Be careful in establishing your priorities. You may wish to list major goals, vital to your business, and include sub-goals that will not make or break your company if they remain unmet. It is likely your sub-goals will increase in number as your business matures.

Strategies. There must be a strategy, or plan of action, to accomplish each goal. For example, under your goal of obtaining sufficient capital to start your business, you will indicate possible sources of start-up money and how you plan to pursue them. These may include your personal savings or funds from sale of property or investments, loans that use property as collateral, or investments by others through private loans or sale of stock in your corporation.

Date of achievement. The dates of certain goals will be fixed if you are to open your business on schedule. Others may be 5 or 10 years down the road. Although dates may undergo revision, include them in your business plan.

Measurement of results. Plan your goals so that you can measure achievement. There's no reason to have a scoreboard if you don't keep score. Not only will you lose motivation, but you may miss out on increased sales volume or new markets if you don't know when you are ready to expand the business.

Modification of Goals

A business is never static. It is either going forward or losing ground. A goal may have been accomplished, or it may need revision or deletion from your list. You will acquire new goals as you and the business mature.

Review your goals and measure results monthly, every three months, or an interval that you decide is appropriate. Don't let a goal nag you. If you have not pursued it and feel it is unimportant, get rid of it. You will find many more to take its place.

Use your sales figures and financial statements to determine the necessity for revision of your goals. Be aware of economic conditions, both locally and nationally.

When more than one person is involved in a business, there should be dates and times set to review and revise goals. Disagreements should be hashed out and decisions made promptly, as soon as necessary data are available. If additional research is required to make a decision, set a time to obtain the information, and set a date to make the decision.

Start-Up and Operating Capital

Capital is the money you will need to start the business and keep it operating. A new business requires not only start-up funds, but sufficient cash to pay operating expenses at least through the first few months.

Your home inspection business will require no inventory, demand little equipment, and result in few accounts receivable (accounts for customers who do not pay at the time of inspection). You must consider, however, the possibility of several months when volume will be low or even negligible.

Your business will need start-up capital, even if you operate out of a corner of your basement. You must determine how much money will be needed, where it will come from, and what it will be used for. This information is a vital part of your business plan.

How Much Do You Need?

Determine how much money will be required by estimating sales volume, start-up costs, and monthly expenses for the first year. Estimated operating and profit ratios and rate of investment return will be calculated from these figures. The estimates can be particularly important in attracting other investors or loan money.

Sales projections. Your market analysis served as the basis for sales projections. You determined the available market, the potential market, and the percentage of both you intend to gain. Estimate your sales, preferably by the month, and multiply those figures by the fee or fees (if you plan to offer varied types of inspections at different fees).

Your projection figures will act as realistic goals and dollar volume expectations, and will assist you in determining the expenses required to meet the volume projections. You may realize that two, not one, additional inspectors will be needed, or perhaps you should choose the larger of two offices available.

Estimated expenses. A listing of one-time start-up costs and monthly expenses should be made. Sample Form "Start-Up Funding" on page 267 indicates the typical expenses a home inspection business will encounter. If you operate from your home, all the expenses will not apply to your business.

The Small Business Administration suggests that businesses should have cash on hand to cover three months of salaries and wages other than the owner's. Use the results of your market analysis, information about your competitors, and personal confidence in your business to make these figures as realistic as possible.

Beyond some initial equipment expense, a new inspector's most important requirement is carrying living expenses while new business is generated. It is not uncommon for a new inspection business to achieve five or ten inspections per week within a year. Many inspectors choose to start part time and build up a clientele before going full time.

Operating and profit ratios. These are obtained by comparison of operating overhead and net profit (before taxes) with the total dollar sales figure for a given period. The ratios can be expressed as actual or projected. In actual operation, the ratios are excellent indicators of your business direction.

Operating ratios in the home inspection industry generally can be estimated at the following percentages of annual gross profits:

√ General management: 5 to 10 percent.
This is what you will pay yourself or someone else to run the company.

√ Advertising and marketing: 5 to 10 percent.
This figure may seem high, but a well-planned and executed marketing policy will build your business.

√ Rent: 1 to 2 percent.

√ Insurance: 1 to 2 percent for general business and liability insurance, 5 percent for errors and omissions insurance, if attainable.
If you self insure errors and omissions, allow 5 percent.

√ Automobile expense: up to 4 percent.
This depends on sales volume, number and type of vehicles.

√ Secretarial: 4 to 6 percent.
This ratio reflects a salaried staff. Contracted secretarial services on a part-time basis generally would result in a lower figure.

√ Telephone: 1 to 2 percent.

√ Office equipment and supplies: 2 to 4 percent.
The higher figure includes computer operations.

√ Credit cards: 1 percent.
An estimated one-third of your business charged to credit cards is reflected in this figure.

√ Bad debts: less than 1 percent.
Although generally paid at time of inspection, fees can remain uncollected and should be reflected in the forecast.

√ Education and training: 1 to 2 percent.
There is a high requirement for skill improvement and keeping abreast of innovations in construction.

√ Accounting and legal fees: 1 to 2 percent.
The cost is relatively low because the accounting methods and issues are fairly clear. This figure does not include any contingency for claims. You'll use the services of an attorney when you are deciding on the structure of the business, and your contract agreement forms. An accountant and an attorney will be kept on retainer, and their services are well worth the cost.

Rate of investment return. This is the percentage of total capital investment represented by net profit for a given period. It is an excellent economic indicator and is especially important to potential stockholders, partners, and loan officers of financial institutions.

Where Will the Money Come From?

You are the first source for your capital requirements. You may have sufficient savings, investments to liquidate, or property to sell. However, don't cut corners too close on your personal assets. It may be wise to consider funds from other sources. These could include:

Partnership. Perhaps another aspiring home inspector and you could pool your capital and set up a partnership. Remember that there will be two of you requiring salaries and/or profits. If your projections don't indicate sufficient sales for two full-time inspectors, a partnership may invite trouble.

You may consider a limited partner who will not contribute to the actual operation of the business, but will furnish a specified amount of capital. Again, remember that the partner will require a percentage of the profits, and you must consider this in your projections.

Corporation. As a corporation, your company can exchange shares of stock for investment by other people, who may or may not share in the actual operation of the business. Most shareholders are uninterested in operating the company. Check with your attorney on all aspects of an incorporated business before you make the decision to use this means of attracting capital.

Business loans. A business loan may just get you over the hump if you lack a small amount of capital, particularly if you wish to maintain a one-person operation. Often a loan with acceptable terms and interest is preferable to sacrificing personal assets and may be a prudent financial step. If your business plan calls for expansion, loans are almost certain to be a part of your future.

Be careful about the terms of the loan and even more careful about the collateral you offer. If you are starting business on the proverbial shoestring, a bank may require that you put up your home or other important asset as collateral. All of your economic indicators could predict success, but if the economy suddenly takes a big dip or the bottom drops out of the housing market, you may find yourself living in your pickup, if you didn't put that up for collateral, too. Better to wait a year or two than to risk losing everything you have.

What Will You Use the Money For?

This should be the easy question, because you already have your start-up costs and operating expenses for the first year. You may have to break down the figures or work them out by month if you plan to use your projection to seek outside funds. You would see a different view of your operation if you calculate the percentage of total start-up money to be applied to each expense projection. This also may be a requirement for loans or other sources of capital.

Budget Forecasts

Now that you are an expert at projections and estimated expenses, are you ready to establish formal operating budgets?

The budgets, by month for the first year, and yearly for the first and second years, should follow the format of financial statements your company will use to reflect actual business operations after start-up.

Sample operating budgets for different levels of volume can be found on page 265.

Budgets vs. Actual Operations

As the business proceeds, the budgets will be compared with actual experience. Your business plan should include comparison procedures and indicate the periods for comparison: by month or quarter, semiannually, or annually. For the first year, monthly or quarterly reviews would be preferable. Later, you may wish to use an annual comparison.

You should provide for budget revision based on operating experience. If there are frequent discrepancies between projected and actual figures, determine if the budgets are out of line, or if there is a need to review company procedures in dealing with expenses or other items.

Choosing Professional Assistance

Whether you are a one-person operation, or president of a corporation that employs 50 people, you will require professional assistance in several fields. Some of these will be optional, but others are vital to the operation of any business, such as legal, banking, insurance, and (usually) accounting.

Choosing a professional is an important decision. The person or firm should have excellent qualifications and references. Check with business colleagues and friends, or with professional associations. Employ professionals familiar with the home inspection business, if possible. If the industry is fairly new in your area, you may need to brief the people you choose as to its concepts and operations.

Include in your business plan the names and professions of those you employ, the operations they will handle for your company, their fees or charges, and an estimate of total monthly expenses for the services of each professional.

Legal counsel. An attorney is a necessary professional for a business operation. Managing a home inspection company without legal consultation could be as dangerous as piloting an airplane without flying lessons.

You will require information about laws and regulations governing the start-up and operation of your business. If you incorporate, the attorney will handle this for you. It would be unwise to use a contract form that had not been approved by your attorney. The first person you will call in the event of a liability claim will be your attorney.

Banking. You will require banking and probably loan services. You need a separate checking account for your business, regardless of its size or type of operation.

Generally, it is preferable to do all of your banking business with one institution at start-up. If you expand later, you may wish to divide your working cash, payroll, and investment accounts among several institutions, but for now you probably will be better off dealing with one bank. You will get to know the bank's officers and personnel, and they will get to know you.

Banking services and types of accounts vary widely. If you are not familiar with these differences, a survey of the local financial institutions will be helpful. Your visits also will give you the opportunity to interview banking officers and learn about the bank's attitude toward small businesses, its credit policies, and its community awareness and participation.

If you have sufficient capital for start-up and do not require a loan, check out the bank's loan services anyway. The necessity for a loan may develop later, and you should know about the bank's loan policies, particularly in regard to small businesses, before you decide to establish your checking account there.

Insurance. Select an insurance company or agent with as much care as you do an attorney or bank. Does the agency have a reputation for professional competence? Can it supply appropriate coverage at competitive rates?

Major types of insurance to consider include:

- General liability
- Automobile coverage for company vehicles
- Fire and theft
- Workers compensation
- Business interruption
- Group life and health
- Business life on key personnel
- Errors and omissions insurance (E&O)

You may have difficulty obtaining the E&O coverage. Many carriers do not handle it, and it is expensive. If you cannot get this coverage, your company should set aside a certain amount monthly in a separate investment account as self-insurance against potential liability claims.

Accounting. You may believe you do not require the services of a professional accountant. This may be true if you have a good working knowledge of record-keeping, business accounting practices, federal, state and local tax laws and requirements, or if there is someone in your family who can assist in this area.

Before making your decision about hiring an accountant, however, consider the following:

Will you have time to perform home inspections and also keep the books? Record-keeping can become a burden and may keep you from other important aspects of your business, such as selling and public relations.

Do you know what records are necessary, as well as others that could be extremely helpful in analysis of your business operation? Adequate and up-to-date records are required to substantiate your tax returns and loan applications, but their most important function is to help you increase your profits.

If you employ other people, are you knowledgeable about payroll taxes and deductions? Do you know what forms to file with federal, state, and local governments and when they must be filed?

A professional accountant may be as valuable to your operation as your attorney, and save you from some of the pitfalls encountered by small businesses. Not only will the accountant be up to date on tax laws and have the necessary forms at hand, but your financial records will be readily available. The accountant can alert you to situations that require your review, from cash shortfalls to increases or decreases in profit margin and return on investment.

Marketing, public relations, and advertising. You may wish to hire a marketing consultant, public relations firm, or advertising agency. Your choice of any of these may depend on the size of your operation, your expansion plans, how your competition uses them, the percentage of the market you are seeking, your need to penetrate new markets, or your own feelings of inadequacy in these areas.

Do your homework on a firm's background and references before you consider using it or signing a long-term contract. Don't succumb to high-pressure sales tactics by some members of the marketing and advertising industries who call themselves professionals.

Remember, too, that you will be paying for some services that may be available at no cost to your company. There may be no extra charge for the newspaper to lay out your ad, for the radio station to write and tape your spot, for the television station to film your commercial. Of course, you must take the time to deal with the media sales representatives yourself, rather than having an advertising agency handle details for you.

The same concern for the use of your time would apply when you consider research on potential markets in adjacent areas, or attempt to expand or upgrade your clientele. It may be cost effective to hire someone to perform these types of service.

You may not require assistance in these areas at start-up, but plans may indicate a need later. If you do background checks on marketing, public relations, and advertising firms while you are researching other professionals, you will have the information available when you are ready for expansion.

Secretarial/bookkeeping services. The operation of secretarial firms is as varied as the operation of home inspection companies. They range from a modern office with computerized equipment to a one-person business with a typewriter on the kitchen table. Check fee schedules as well as experience and background. It may be more important to save money on office expenses than to save reports on a computer.

If there is no one in your office during business hours, it is advisable to contract with a telephone answering service. Many potential customers could be turned away by your taped telephone message to leave their name and number so you can call back. Expensive, high-volume advertising may be a waste of money if there is no real person at the telephone number listed in the advertisement. Most telephone answering services today feature electronic equipment to keep you in constant touch with your business. One interesting option is a cellular phone in your car with call forwarding from the answering service.

Bookkeeping services are as varied as secretarial services. Be certain the bookkeeper has the expertise to handle your requirements in an accurate and timely manner. Some firms combine secretarial and bookkeeping functions. This could be feasible, provided it furnishes competent personnel in both areas.

Laws and Regulations

Federal, state, and local laws and regulations will affect your business. It is your responsibility to be aware of and follow these laws and regulations. If you require assistance in obtaining this information, your attorney may be the proper person to contact.

The following requirements may be applicable to your business. If so, your methods of complying with them should be included in the business plan.

Licensing

The only state presently requiring licensing of home inspectors is Texas. However, several other states are considering licensing laws. If your state is one of them, you and your inspectors, both salaried and contracted, should be preparing for the test that probably will be required for a license.

Consumer and Environment Protection

There are federal and state laws, and local ordinances, with which you should be familiar. Home inspectors who investigate environmental hazards would come under the regulations of the federal Environmental Protection Agency, its state counterpart, and perhaps county and city agencies. All the states and many local governments provide protection to consumers in various ways.

Labor Relations

If your company employs a salaried staff, you will need information about federal legislation, including the Fair Labor Standards Act, Occupational Safety and Health Act (OSHA), and Civil Rights Act, and state laws governing workers compensation and fair employment practices.

Taxes

Taxes and tax law requirements are a major headache for the business person. The Small Business Administration publishes a yearly booklet titled "Tax Guide for Small Business," prepared by the Internal Revenue Service. The information includes social security taxes for owners and employees, withholding requirements, corporate income taxes, personal income taxes of sole proprietors and business partners.

State and local governments impose income, property, sales, business license, and unemployment compensation taxes, among others. State and local government offices can furnish information, but it may be helpful to discuss your tax requirements with your attorney and accountant.

See page 270 for a sample worksheet to list your tax obligations, due dates, where paid, and date paid.

Incorporation

Each state has laws regulating incorporation procedures and operation. Your attorney can advise you on the advantages and disadvantages of the corporate business format, as well as the requirements of your state.

Corporate income tax returns can be complicated and should be handled by your accountant.

The Business Office

This section of your business plan includes a description of the office, management and staffing requirements and procedures, legal requirements for the office, and maintenance of financial records.

Consider the following items in describing the establishment and operation of your company's headquarters:

Location

Description:
- Site
- Furnishings
- Equipment and supplies

Management:
If more than one person is involved in management of the office, establish duties and responsibilities of each. List departments and/or staff positions. Be sure to specify who will sign company checks.

Legal Requirements:
- Compliance with federal, state, and local laws and regulations
- Incorporation procedures, if applicable

Financial Records:
Establish bookkeeping procedures and set up the books. Assign responsibility for the financial records.

Staff:
- Organization chart
- Job description and salary range for each position
- Working hours, employee benefits (insurance, vacation, sick leave)
- Personnel training and supervision methods

Overhead Expenses:
Determine procedures for comparison of budget projections with actual expenses. Monthly comparisons of actual expenses also will be helpful.

Company Policies and Procedures

Every company has policies and procedures, written or not. There are methods of operation you will apply, and decisions you will make regarding fees, inspections, personnel, and other facets of the business.

Work out these methods and decisions prior to start-up, even if you are a one-person operation. You save time and energy, maintain consistent business policies, possibly prevent liability claims, and probably operate a more profitable business.

Following is a list of suggested items to include:

Inspection fees. Set up your fee schedule to include single-family homes, condominiums, single units of condominiums, new construction, commercial buildings, partial inspections, environmental hazards or other specialized inspections, and other types of inspections you plan to perform.

Method of payment. If your business is based on fee paid at time of inspection, specify exceptions you will make and how payments will be handled. State the procedure for review of your fee schedule at appropriate intervals and who will make the decision to revise the fees.

Inspection checklist. This will constitute the basis of your home inspection business. It should be referred to in your business plan and a copy attached as an appendix. No home inspection should be performed without a copy in the inspector's possession, whether that inspector is you, an employee, or a subcontractor.

Customer agreements. State your policy regarding pre-inspection agreements and attach a copy of the agreement that you will use. Be specific about use of the agreement, presentation to the customer, situations where the agreement may be waived or changed, and alternative procedures that will be used in such cases.

Inspection reports. Indicate the type of written report you will use: checklist, narrative, or a combination. A sample report should be attached.

Personnel policies. Specific policies will save confusion and indecision hiring and training employees, and help employees to know their functions and how to perform them.

√ Personal standards:
   Identify the personal characteristics you seek and your expectations of the employee's performance.

√ Job title and duties:
   Each position should have a title and job description with the duties specifically outlined.

√ Interviewing and hiring:
   Be specific about the methods you will use, the type of application you will require, and the questions you will ask. Know what you plan to tell applicants about your company, duties of the position, salary, benefits, and performance expectations. Have a definite policy about follow-up with applicants.

√ Training and supervision:
   Indicate who will train and supervise each type of employee and the methods to be used.

√ Evaluation and promotion:
   Have a specific schedule for formal evaluation of employees and stick with it. During the evaluation process, give the employee the opportunity to evaluate you as an employer. How to retain excellent employ-

ees and how to assist average employees to perform at higher levels should be part of your business plan.

√ Termination:
Include termination procedures; notice of termination and the means by which the employee may avoid termination; firing without notice for cause, such as embezzlement or improprieties during inspections; and any financial considerations offered employees terminating voluntarily or being terminated by the company.

Subcontracting procedures. Have specific procedures for hiring and evaluating subcontractors. You may have sales figures or other criteria to apply in your decision to hire subcontractors. Any policy regarding subcontractors should be made a part of the business plan.

√ Hiring standards:
State your policy regarding background and reference checks, other hiring procedures, interviews, contracts, and qualifications of the subcontractors.

√ Amount and method of payment:
Be specific about when a subcontractor will be paid. Indicate the percentage of the fee that will be paid to the subcontractor, the schedule of fees, or a statement that terms will be negotiated.

√ Evaluation:
Indicate evaluation procedures and when you will evaluate subcontractors, either by time period or by number of jobs. The evaluation could include a checklist rating on the following:

- Quality of work
- Appointment scheduling and promptness of service
- Neatness of appearance and courtesy
- Fee schedules, if applicable, and expansion capability
- Loyalty to your company
- Customer relations

A number of home inspection companies require new inspectors to sign a non-compete agreement before they are accepted and trained. The reason frequently given is that this is a personal contact business, and most good inspectors get more requests than they can handle.

On the other hand, if you pay your inspectors fairly they are unlikely to leave. And if they do, there is plenty of business for everyone. In many states, a non-compete agreement is not legally enforceable.

Operations manual. If you employ several home inspectors, it may be advisable to write an operations manual for field personnel. The business plan would indicate how it is to be used, and the company policies included in the manual. A copy should be attached.

Company records. You may wish to list formal records that will be maintained. Your business plan should indicate:

- √ Title of each record
- √ Category: client, subcontractor, personnel, equipment, bookkeeping, etc.
- √ Personnel responsible for maintaining the records in each category
- √ Personnel responsible for collection and dissemination of information in each category through reports to management

Review of company policies and procedures. Policies and procedures should be brought up to date periodically. A policy may outgrow its usefulness, or expansion of the company may require additional procedures. The business plan should include:

- √ Specific time periods for review
- √ Personnel who will review and recommend additions, deletions, and revisions
- √ Procedures for review
- √ Company officer(s) who will make final decisions on the recommendations

Marketing and Sales

These are not two separate entities. Marketing is an prerequisite for sales; sales are the result of appropriate marketing procedures.

Marketing encompasses everything you do to promote the sales of your services, from your first customer's confidence and respect to the most expensive advertising campaign money can buy.

The business plan should detail procedures for marketing your services. Through your research, market analysis, and sales projections, you should have a clear concept of the types and locations of customers you will seek and the sales methods you will use.

Logo

A logo is a printed symbol of your company, its name, and reputation. It should be simple, well executed, and easy to recognize. Many commercial artists and advertising agencies specialize in the creation of logos for new companies.

Your logo should appear on every piece of printed material and be part of every applicable form of marketing and advertising:

- Business cards, stationery, invoices, contracts, report forms
- Office sign and business vehicles
- Brochures and press release forms
- Printed advertisements: newspaper, magazine, Yellow Pages, direct mail, giveaways
- Television commercials in conjunction with the name of your company on the screen

A company slogan can be used along with a logo. It has the advantage of being usable for radio spots, as well as printed materials. There are thousands of excellent business slogans, but be careful about wording; usually the shorter, the better. Slogans often can be "corny" or have unanticipated results. Sometimes a humorous slogan helps the public remember a company with affection; however, this may not be true with a home inspection service. Give a slogan a thorough test run on family, friends, and acquaintances before deciding to use it.

Advertising Media

Advertising takes many forms. Consider the effect each might have on the acceptability and prestige of your company.

Be selective in your advertising approach. Establish a strict budget and learn to say no. The day you open your door to business, and probably much sooner than that, you will have advertising account representatives coming out of the woodwork. You may not see a customer during your first two weeks in business, but you won't be lonely. Everyone will try to sell you something. Keep a tight rein on the checkbook and follow your business plan.

You may wish to consider some of the following forms of advertising:

√ Newspapers and magazines, especially local community publications

√ Radio and television

√ Yellow Pages of the telephone book

√ Direct mail, including brochures, letters to potential customers, discount coupon books, newsletters

√ Giveaways, such as pens, calendars, hats, mugs. If your logo lends itself to three-dimension format, an excellent giveaway would be a key chain or other item in the shape of the logo.

Public Relations

Your best public relations expert is yourself, and your number one asset is the reputation of your company. You have already contacted business, government, and civic leaders in setting up your company. The business plan should describe how you will continue your public relations efforts. Establish a schedule of calls on customer sources, the media and civic organizations.

If your talents run more toward management or actual inspection work, you may have an employee handle public relations or contract with a public relations firm. Another possibility could be someone you know, such as a retiree looking for a part-time position, who is a good speaker and interested in the positive effects of the home inspection industry. A limited partner or someone on your corporation's board of directors are other possibilities.

Public relations endeavors may include:

Personal calls on real estate brokers, mortgage officers, local building inspectors (if government inspections are contracted out), real estate and trust attorneys, newspaper business editors, and public service directors of radio and television stations.

Membership in local trade associations and civic organizations.

Presentations to real estate and mortgage banker associations, local chambers of commerce and other business groups, and fraternal and civic organizations.

News releases (typed on company stationery or news release forms) transmitting feature articles and public service announcements about home inspections, maintenance, and safety to local newspapers, newsletters, and radio and television stations.

Participation in home and trade shows, community festivals and fairs, and other organizational exhibits.

Record Keeping

Business records maintained in a timely and proper order are essential to the efficient operation of a company. Management should have immediate access to any record at any time.

Whether a record-keeping system is maintained manually or computerized, it should be up-to-date and accurate. Your business plan should describe the record system and its components, how it will be maintained, and the person(s) responsible for management of the system.

The following is a list of essential records in any home inspection operation. Your company's particular needs may require additional records:

Customer Records

Each customer file should contain:

√ Full name, home and business addresses, telephone numbers, type of referral

√ Copy of pre-inspection agreement

√ Copy of inspection report and record of transmittal time and method

√ Backup material and notes, including photographs if used

Bookkeeping Records

√ Cash journal to record cash receipts and disbursements, and petty cash journal if applicable

√ General ledger and classification of accounts
- Income
- Expenses
- Assets
- Liabilities
- Equity (net worth)

√ Separate ledgers for certain accounts, as determined by management
- Sales
- Accounts payable
- Taxes payable
- Accounts receivable, if applicable
- Owner equity
- Certain types of assets, including depreciation account for each depreciable asset

√ Profit and loss statement (statement of income and expenses)

√ Balance sheet (statement of assets, liabilities, and owner equity

√ Worksheet tax record for federal, state, and local taxes (type of tax, due date, when paid, amount paid)

Samples of several of these records will be found at the end of this chapter.

Subcontractor Home Inspector Records

Each subcontractor file should contain:

√ Full name, address, telephone number, business identification number or social security number (for company or individual, respectively)

√ Qualifications, background check, references

√ Assignments, including:
- Job site and name of customer
- Date assigned
- Date completed
- Fee
- Date of payment

√ Periodic evaluations of subcontractor

The file should also contain documentation of oral or written comments from customers.

Personnel Records

Each personnel file should contain:

√ Full name, address, home telephone number, social security number

√ Application and/or resume

√ Job title and description

√ Employee's withholding certificate (IRS Form W-4)

√ Wage rate and benefits

√ Hours worked and reimbursable expenses for each pay period

✓ Payroll record
- Earnings
- Deductions

✓ Job performance
- Evaluations
- Wage increases
- Promotions
- Other pertinent data related to performance or evaluation

✓ Termination
- Type
- Reason

Summary of Business Plan

DESCRIPTION OF THE BUSINESS
- Title of the company
- Location
- Operation
- Purpose

BUSINESS PHILOSOPHY AND OBJECTIVES

MANAGING THE BUSINESS
- In the field
- In the office

MARKET ANALYSIS
- Description and methods
- Researching the competition
- Your market approach

BUSINESS GOALS AND STRATEGIES
- Determining the goals
- Writing the goals
    - Priorities, strategies
    - Projected achievement dates
    - Measurement of results
- Modification of goals

START-UP AND OPERATING CAPITAL
- How much do you need?
- Where will it come from?
- What will it be used for?
- Budget forecasts
- Budget versus actual operations

CHOOSING PROFESSIONAL ASSISTANCE
- Legal
- Banking
- Insurance
- Accounting
- Marketing, public relations, advertising
- Secretarial and bookkeeping services

LAWS AND REGULATIONS
- Licensing
- Consumer and environment protection
- Labor relations
- Taxes
- Incorporation

THE BUSINESS OFFICE
- Location
- Description
- Management and staff
- Legal requirements
- Financial records
- Overhead expenses

COMPANY POLICIES AND PROCEDURES
- Inspection fees
- Method of payment
- Inspection checklist
- Customer contracts
- Inspection reports
- Personnel
- Subcontracting procedures
- Operations manual
- Company records
- Review and revision of policies and procedures

MARKETING AND SALES
- Logo
- Advertising
- Public relations

RECORD KEEPING
- Customers
- Bookkeeping
- Subcontractor home inspectors
- Personnel

8. OFFICE PROCEDURES

Introduction

Effective office procedures can mean the difference between satisfied and dissatisfied customers and can have a direct impact on your profit and loss statement.

Whether you operate from a desk in the corner of your den or a fully automated, plush high-rise office with a staff of 10 people, it is important to establish office procedures and follow them. Your customers will be happier, your employees more efficient, and your records more accurate.

Your business will be an entity with form as well as content. The content is your knowledge and the services you deliver; the form is the methods you use to deliver the content. Your company is content fleshed over a skeleton of form. Office procedures are the skeleton.

This chapter begins with the tasks of that key employee, your secretary/office manager, and ends with the owner's role in office procedures. While you must take into account the customer's perception of satisfaction, it is the wise owner who values and motivates employees. You are assuming the leadership position, but it is your organization that spells success, not just your own dynamic personality shining on the world at large.

We're in the communications business. This means we must communicate internally as well as externally. We must have a performance standard for every employee and subcontractor, as well as for ourselves. By adhering to the standard, we set the example which becomes a touchstone for internal communication and the motivation of our employees and co-workers.

This boils down to your acknowledgement of the employees' perception of fairness: reasonable expectations, comfortable surroundings, true appreciation, and fair and predictable rewards. By creating these conditions for your internal business family, you will establish a sense of comradeship and loyalty which will be reflected in the relationships between your company and the marketplace.

The main instrument for creating the sense of predictability in your company operation is a Company Policy Manual. An sample is included in this chapter.

Being fair, reasonable, honest, and accommodating are more important than being the least expensive. It is important to be fast as well as accurate, but if you build a company where people are valued, where loyalty flows in both directions, and where the unusual need will be met because people care and the company is a motivated team -- you will have captured the secret of success.

Belief in your organization, your people, your policies, and your mission are key ingredients. Home inspection is not an exact science, and the understanding which can find the facts suggested by your observations, must also report those facts in a context that is useful to the customer.

You're in a business of the heart as well as the head, and the way you regard your own people and practices will come back to you, for it will define the quality of your company beyond the product you deliver. The public will recognize this, and you'll either win or lose on attitude.

To demonstrate a method of following up a call from a potential customer, through the sale, inspection and report, to payment of the subcontractor and proper disposition of customer documents, this section will set up a hypothetical office. The company is owner-operated, hires several subcontractors, and employs one full-time secretary/office manager who handles the small office.

Collect Customer Data

Your response to a call from a customer should reflect professionalism and concern for the customer as an individual, not just a potential sale. Whether you use a telephone answering service, employ a secretary/office manager, or have an answering machine, be certain that the call is received courteously and your company name is clearly stated.

The person answering the call fills in a pre-printed form, or job sheet, while obtaining information from a customer during the telephone conversation.

The job sheet will include:

√ Customer's name, address, home and office telephone numbers

√ Scheduled date and time of the inspection

√ Property information
   • Address and general location
   • Directions to the site
   • Selling price
   • Name of present owner
   • Type of property (new, used, or remodeled home; apartment building, condominium, commercial; owner-occupied, tenant-occupied, or vacant)
   • Special instructions (partial inspection, additional inspection items, close-in construction)

√ Real estate agent (name, company, address, office and home telephone numbers

√ Inspection Fee
   • Amount
   • How paid
   • Credit card information, if applicable (type of card, account number, expiration date)

√ Miscellaneous information
   • Who will be present at inspection
   • Referral source
   • Name of person making the call if other than the customer
   • Name of employee completing the job sheet, date and time
   • Name of inspector assigned to the inspection

As the customer provides information, the person taking the call answers questions about the home inspection, its importance, what is and is not included. Be clear about the fee and time of payment, the use of a pre-inspection agreement if applicable, and other company policies.

If the customer has questions that cannot be answered immediately, or if there are unusual aspects of the inspection, the owner or assigned inspector should return the customer's call as soon as possible.

Scheduling

One copy of the job sheet is attached to the job board in the office. If the company has more than one inspector, there will be a slot for each one.

A second copy will become the initial document in the customer file, established when the appointment is scheduled.

The secretary/office manager schedules inspection appointments and maintains an appointment calendar for each inspector.

Subcontractors

The secretary/office manager maintains a file for each subcontractor hired by the company. If they work on a continuing basis, they visit the office each evening to submit the day's inspection reports and check the job board for new assignments.

If the subcontractors are not regulars, contact them by telephone as they are needed and, if necessary, clear appointments with them.

Each subcontractor file includes a weekly report of scheduled assignments, completion dates, inspection fees, and date of payment. You should receive a weekly inspection recap, including the total amount due each subcontractor. The secretary/office manager may write the subcontractors' checks for your signature (in some offices, this function is handled by the owner, accountant, or bookkeeping service).

The secretary/office manager is responsible for transmittal, in person or by mail, of each contractor's weekly report and payment check. Each subcontractor is also furnished with a supply of necessary forms, such as checklists, contracts, reports, and other documents.

The Checklist

A checklist is a necessary tool for every home inspection operation, and it is essential for a company that hires subcontractors. Standards of Practice, your own or the ASHI version, should be duplicated so that each inspector has several copies at all times. Then there will be no question what procedure the inspector in the field is required to follow.

Whether you use the HomeTech BAR or another version of a checklist, the printed guide does several important things:

✓ An on-site reminder to the inspector to make a complete inspection

✓ The nucleus of the final report to the customer, whether a copy is used as the actual report or the information is the basis for a narrative report

✓ An essential component of the customer file, describing each aspect of the inspection and how it was handled

✓ An aid in the event of a liability claim, especially if the inspector notes unusual conditions or circumstances

A report form published by HomeTech can be utilized as a combination contract agreement, checklist, and final report. The form, titled "Building Analysis Report," [BAR] is comprehensive and includes a total of nine separate pages printed front and back.

Each page consists of two copies. The original, printed on carbonless paper, can serve as the final report to the customer. On the back of each page, the customer will find helpful information about home construction, suggestions for home care, and estimated replacement costs of many building components and systems. A lightweight cardboard copy is intended for the home inspector's company file.

The cover page of the BAR contains the pre-inspection agreement, and includes the following:

✓ Statement of the purpose of the inspection

✓ Description of items to be covered in the report

✓ Statement of what the inspection does not cover:
    Latent and concealed defects and deficiencies
    Governmental codes and regulations
    Present and potential environmental hazards
    Instruction on home maintenance
    Dismantling of equipment, items and systems
    Swimming pools, wells, security systems, and other items
    Presence of rodents, termites, or other insects

✓ Liability disclaimers regarding the inspection and report

It is the home inspector's responsibility to make sure the client reads the pre-inspection agreement, understands it, and signs the agreement.

At the bottom of the cover page there is space for recording fee payment, which serves as a receipt for the customer. The inspector fills in this information and collects a check at the time the report is conveyed.

The Report

How the inspection report is prepared and transmitted to the customer is determined by company policy or by specific customer preferences. The two major delivery systems are the written report and the site report.

As a general practice, a written report will be prepared and conveyed (mailed, made available for pick-up, or delivered) within 48 hours of the inspection. It is easier and more profitable to deliver an immediate report at the inspection site.

This is by far the fastest, easiest, and least costly way to operate the business because you (the inspector) are handed a check when you deliver the report. With a little practice you or your inspectors can easily become proficient in the use of the BAR. Purchasers and real estate agents do not want to wait for the inspection results, and prefer to get the report on the spot.

Timeliness of the report is an important factor, because an offer to purchase or negotiation of a sales figure may be riding on your inspection report. The customer has paid the inspection fee and is anxious. A primary responsibility of the home inspection company is to provide the report on time.

There are several ways to handle the report. They include the following:

The inspector dictates a narrative report on tape en route to the next appointment or back to the office. (This may or may not be a good idea, depending on the hazards of traffic in your area!) Cassette tapes are transcribed into final form and prepared for the inspector's signature.

You can avoid narrative reports by using a good checklist with ample space for notes (HomeTech BAR). By spending a few minutes at the site to complete the form and explain items to the client, you establish excellent communication without paying for administrative work back in the office.

Using a standard format, the inspector fills in blanks or dictates supplemental information, crosses out extraneous material in the form, and completes drafting a final report ready for typing. The client may receive a summary report at the site and wait up to 48 hours for the final report.

If the company has more than one inspector and everyone keeps fairly busy, total report volume could overwork a single employee. A word processor with ample memory or disk storage may solve the problem without going to an expensive computer system. A pre-formatted letter is on file and ready for written or dictated completion.

Regardless of the method chosen, be consistent in following established policy for an efficient and cost-effective office operation. Constantly switching methods may lead to confusion and error. You can be flexible as circumstances require, but each variation should be clearly understood by every member of your staff.

Speed is a prime requirement, but accuracy is a necessity. Time should be allowed for thorough proofreading and double-checking all figures. This is particularly important if outside secretarial services are used. If a report reads $500 when it should have read $5,000, the company could be in serious trouble.

Especially when the company has more than one inspector, time constraints usually require the staff to proofread reports. This is not ideal, but HomeTech's experience over 20 years has been good, with no major claims because of errors in a report.

A cover letter accompanies each report that is mailed. The customer is thanked for the business, and if possible a personal note of some kind is included. A well-written, easily personalized cover letter stored in a word processor's memory can be an effective sales tool.

The Customer Questionnaire

A final document in the customer file is a questionnaire submitted to the customer following the home inspection. Not only do you obtain important feedback about your company as perceived by the customer, but the questionnaire also is a useful public relations tool.

Your company must meet the customer's perception of satisfaction, not yours. A satisfied customer may tell three or four friends about your company, but a dissatisfied customer probably will tell 15 or 20 people.

The questionnaire gives the customer an opportunity to let you know of any dissatisfaction. A follow-up call to offer additional information or straighten out a minor problem may turn around the customer's perception of your company.

The questionnaire should be worded to avoid yes or no answers. Several possible answers are offered, with space for a check mark by the chosen answer.

Sample questions and answers include:

Professional manner of inspector:
[ ] Very professional
[ ] Professional, but left some doubt in my mind
[ ] Did not project a professional image

Ability of inspector to communicate:
[ ] Inspector communicated well
[ ] Inspector seemed interested in my questions but appeared to be vague
    about certain items
[ ] Inspector seemed to be in a hurry and did not answer all the questions
[ ] Inspector showed little interest in communication

Other questions may relate to:

- Overall satisfaction of service
- Promptness of inspector
- Technical expertise of inspector
- Thoroughness of inspection
- Promptness of written report
- Willingness of inspector to follow up

Referral Source Questionnaire

A questionnaire directed to your referral sources is another useful sales tool and quality control method. Real estate companies who regularly refer customers to you are logical targets for the questionnaire.

It also may be interesting to send a questionnaire to companies that have made only one or two referrals or have not made any for some time. You may learn more here. The questionnaire is also a good promotional aid in attracting attention to your company.

The referral source questionnaire follows the same format as a customer questionnaire. The wording would be somewhat different, and there are additional questions you can ask.

✓ Have you used our company in the past 12 months?

✓ If so, approximately how many times?

✓ Do you recommend home inspections to your clients on a regular basis?

✓ If so, do you recommend one specific home inspection firm more than others?

✓ Do you regularly accompany the home inspector and the purchaser during the inspection?

You may also want to list the names of your inspectors and ask which have performed inspections for the real estate agent's clients. An evaluation of each inspector, based on a multiple choice of Excellent, Good, Fair, Poor, with space provided for comments, is helpful.

Another effective question would compare your company with other home inspection companies in the following areas:

- Items covered in the inspection
- Cost of inspection
- Apparent qualifications of inspectors
- Promptness of inspectors
- Promptness of written reports to buyers
- Ability to schedule within your needs
- Overall satisfaction

Fees

Fees are the lifeblood of your business and should be handled with care. You may say, "Why worry? I've got it made. Fees are paid up front and I don't have to worry about inventory, cost overruns, and petty theft."

True, but in addition to the fact that your fees should be determined by market analysis, sales projections, and cost estimates, you must collect them in a professional manner.

The person who takes the call and sets up the appointment must be specific in stating the amount of the fee and the condition of payment. There should be no question in the customer's mind about when the fee is due.

Fee Collection

Some people are very good at deferring almost any kind of payment and making you believe they have a good reason for it. Your policy on immediate payment at time of inspection should be firm.

There will be times when the customer is out of town and must have the inspection completed at once. Your policy may be to inform the customer that only the owner can waive the rule, and have you return the call. It is professional practice to require a retainer before work is undertaken, so you need feel no hesitation about being paid before you invest staff time and effort on a new client.

If the customer will use a credit card, the job sheet will contain the information necessary to verify the account. When payment is by check, HomeTech's experience indicates that there is no need to request a certified check, or to ask for additional identification. Bad checks are extremely rare.

If a check does bounce, a phone call to the purchaser is usually all it takes to make it good. The phone call should be low key, professional, and give the client no reason to take offense at your call.

Accounts Receivable

There is no law inscribed in stone that home inspection companies must collect at the inspection site. It just seems to work better that way, especially since you do not serve the same customers on a continuing basis. Most home inspectors have found no difficulty in collecting fees at that time.

Occasionally it will be suggested that the inspection fee be included in the settlement funds on a property. You are likely to have two problems with this: your payment is help up for 30 to 60 days, and if the sale does not go through you can expect problems collecting.

You may be willing to allow your customers to defer payment, or you may have a few accounts receivable through no choice of your own. Maintain an accounts receivable ledger to provide up-to-date information on overdue accounts.

Some people are very efficient (intimidating) in collection work, and you or your secretary/office manager may prove to be one of these. However, as a last resort, you may have to deal with a collection agency, have your attorney handle the collections, or write off the accounts as bad debts.

Office Overhead

Office procedures as outlined in this chapter have been based on a small company's operation which includes owner management, subcontractor inspectors, and a non-automated office with one staff member.

There are other options for a small company. In most home inspection companies the owner is the primary inspector, with income derived primarily for doing inspections. Your market analyses and sales projections will influence your decision in this area.

If you do not believe your company could withstand the overhead costs of a separate office and a full-time staff member in its initial months of operation, consider the following methods of handling calls, paperwork, bookkeeping, and other office duties.

An office in your home. This would include the part-time assistance of a family member, answering machine, and/or answering service to handle calls, a secretarial/bookkeeping service to type reports and keep the books, a typing service to complete inspection reports, and/or other part-time help.

A separate office that is not staffed. When you are out, you would handle incoming calls as noted above and use similar methods for your reports and bookkeeping.

Upgrading the Office Operation

Your sales projections and capital funding may offer the opportunity to start your company on a larger scale. Or your projections may indicate a possible upgrading within the first year or two.

Following are examples of office and staffing that may fit your operations:

1. Two or more salaried or subcontractor inspectors doing most of the field work; secretary or receptionist to handle incoming calls, typing, and filing in a non-automated office; owner acting as office manager, field superintendent, sales manager, and public relations representative.

2. Automated office procedures; sales and public relations handled by an office manager; secretary/receptionist; owner and subcontractors in the field. Computer operation in schedules, contracts, checklists, inspection reports, subcontractor records, generation of mailing lists and promotional material, and accounting procedures and documents.

3. Three or more salaried or subcontractor inspectors; fully automated office procedures; owner acts as office, sales, public relations manager; office staff of two persons who handle all calls, computer operations, and bookkeeping functions. Field work is supervised from the office by senior office employee; field communications and reporting handled electronically through computer modems.

Determining Your Office Needs

Primary considerations in determining your office needs include:

- The available market for your services
- Sales projections
- Estimated overhead
- How much start-up capital you are willing to risk on office overhead
- How much you enjoy attending to details and paperwork

If your market analysis indicates customers are out there waiting for you, and you believe you can hire several competent subcontractors, and your talents run to promotion and public relations, you may decide to set up a fully staffed, automated office at the very beginning.

Or you may be the type who doesn't like to take risks of any kind, so you'll just set up your desk in a corner of the den nobody uses anyway because they're all watching TV in the family room. You know the market is out there, but you have decided to start slowly and see what happens. Besides, you like to do the actual inspections, and you don't believe in all that public relations hype.

At either extreme or anywhere in between, you must analyze yourself and your estimated costs before you can make an intelligent decision about the kind of office you need, or don't need.

If you have worked up reliable market analysis and sales projection figures, you must base your office requirements on these figures. Mr. Super Salesman may not need two secretaries and a fancy computer system, but Mr. Modest may not be able to handle a profitable business alone.

You could calculate monthly office overhead on two or three different office setups, if you wish. Your overhead estimates will include figures for the following items:

- Office staff (salaries and benefits), including salary of owner
- Rent and utilities
- Office equipment and supplies
- Telephone and answering service
- Advertising
- Insurance (including unemployment compensation and medical benefits)
- Taxes (including employer-paid FICA)
- Professional services (legal, accounting, secretarial, telephone answering)
- Education and training

With your estimated monthly office overhead expenses for each type of office setup, multiply each figure by 12 and compare the totals with estimated annual sales figures.

Your calculations are not complete until you decide how much you are worth to the company. It may be feasible for Mr. Super Salesman to hire people to mind the office and oversee the field work, or for Mr. Modest to hire a salesman and concentrate on the field work and office chores.

Don't forget the paperwork. If you don't like to do it, you have two choices:

1. Discipline yourself to do it -- even if it means working weekends

2. Hire someone to do it -- which could be a more cost-effective solution in the long run

One of the best features of the home inspection business is that paperwork is kept to a minimum, especially if you use an on-site report such as the HomeTech BAR. The main advantage to having any office staff is that inspection appointments can be confirmed on the spot, so that agents and purchasers need not wait for a return call. In fact, nearly 75% of all inspection companies today function without a full time staff person.

It is risky to employ a full time salesperson to market your organization unless you are **very** well financed. It is highly unlikely that volume can be built up quickly, regardless of the effort. What is more, while a sales representative can make calls and set up real estate sales meetings, the actual presentations must be done by the home inspector. Real estate agents want to meet and talk with the actual inspector, not a representative.

Try to determine the positions required to handle the business generated by your company. Don't forget that you may be willing to work 16-hour days, but your employees generally will limit their workdays to 8 hours. You can now match the office setup to your business, the money available, and your talents as owner.

Office Records

There are four essential financial records that should be maintained in the company office:

√ Cash journal (daily)
  • Cash received (fees, accounts receivable, other)
  • Cash disbursed (salaries, subcontractor fees, and other expenses)

√ Sales report (monthly)
  • Leads
  • Scheduled appointments
  • Inspections
      - Location by area
      - Fee
      - Inspector
      - Requests

√ Production report (monthly)
  • Total volume
      • By area
      • By inspector

√ Financial statement (profit and loss)

Other financial records may be maintained in the office or by a bookkeeping service or accountant: general ledger, accounts receivable, accounts payable, payroll, taxes, balance sheet, and any others required by management.

Customer, personnel, and subcontractor files also are maintained by office staff. In addition to the schedule book, a telephone log is usually kept which can serve as a sales lead record.

If the office is automated, there are numerous additional records to assist sales and public relations efforts. These will be discussed in the next section.

Automated Office Operations

At start-up, your company may be a small or one-person operation. You may pick up messages from an answering machine or a secretarial service. You may keep the records in manual form or use a computer for records and reports.

If your home inspection operation will be part of an existing business, you may be using automated equipment now and wish to adapt your new operation to it. There is software available for complete automation of a home inspection business, from the incoming call to preparation of the written report, to cover letter and envelope, and can include direct transmittal of a subcontractor's report to your computer.

The Computer Decision

A sizable investment in computer equipment at start-up may be a mistake. In addition to the expense involved, you may be required to spend more time than you care to learning the computer and how it can help your business. You may be lost in the maze of available software. If you plan to use checklist reporting, a computer system may not be cost or time effective.

Many home inspectors postpone computer investment until they have several months of operational experience to use as a basis for intelligent decisions. Since inspection fees are usually collected at time of service, book-keeping is fairly simple and may not require automation for efficiency and accuracy.

If you plan to expand your business, however, and intend to automate office procedures at that time, it is preferable to install the computer system well in advance of the projected expansion. The more training and adjustment time you allow, the smoother the transition will be.

With the right system, software, and personnel, a computer installation could handle every office procedure. You will draw the line, hopefully, at telephone answering. You may safely assume that no one, especially a potential customer, appreciates talking to a computer over the phone.

The computer allows instant access to scheduling, financial reports, customer files, and subcontractor status. In addition to office procedures, an automated system can assist in market analysis, sales promotions, and public relations.

The computer requires only one entry of the name of any customer, subcontractor, lead, ledger account, or referral source. It offers indefinite storage and availability of all information important to your company's operations.

A home inspection business could utilize this information to generate a variety of sales, financial, customer, marketing, and promotional tools. A sample could include:

✓ Computerized mailing lists for promotional material
  • By geographic area
  • By type of customer
  • By referral source

✓ Follow-up to customers and referral sources
  • Thank-you letter
  • Inspection questionnaire

✓ Accounting
  • Bank deposits
  • Journals
  • Ledgers
  • Financial statements
  • Operations and sales reports
    - By individual inspector
    - By type of inspection
    - By amount of fee
    - By referral sources

The Owner's Role in Office Procedures

Your company's office procedures will reflect the type of individual you are, your perceptions of efficient office operations, and the position in which you see yourself as business owner.

In a small company, the owner is usually a "doer" who performs most of the operations.

As the company grows, the owner must become an "organizer" of time and effort expended, by the owner and anyone else in the business.

The next step in growth requires the owner to become a "director." The operation will include full-time personnel, but the owner directs the business by telling the staff what to do and how to do it.

The fourth stage forces the owner to become a "delegator." The organization is too large for one person to operate. The owner must hire people, give the people responsibilities, and then let go. Can you keep hands off and allow someone else to do the job? Even the most dynamic and successful "one-man show" reaches the point where the owner's hands cannot be on every cutting edge. If your enterprise is to succeed, the skills used in delegating responsibilities to good people will become the owner's strongest asset.

Summary

The secretary/office manager is the most important person in a small office operation, who takes the initial call and makes the sale, who collects customer data, schedules inspections, prepares company reports and subcontractor records, generates inspection reports and letters for the customers, and who conducts market research and soothes irritated clients and/or subcontractors. The secretary/office manager is your Very Important Person, whose skills can make a big difference to your company's success. Don't overlook these talents and skills when your company gets to the stage that a full time employee is in your plan.

In addition to your pre-inspection agreement, which is designed to shape customer expectations, protect against liability claims, and trigger the actual inspection, you should use two important questionnaires to help measure your company's performance against the customer's expectations. The customer questionnaire collects information directly from the user of your services, and the referral source questionnaire goes to companies that refer your services. These research aids can give you information to modify your services to reach specific market goals.

Fee collection is normally done at the inspection site or by credit card with the details handled at the time an inspection is scheduled. You may not feel it is necessary to collect all fees at the site, but you will be happier if you do.

You can start very small and work into a fully staffed professional office. A corner of the den or basement, an answering machine, and a secretarial service or the assistance of a family member who can type and do light bookkeeping are all you'll need. The HomeTech BAR includes the pre-inspection agreement and a record of payment with the report.

If your market analysis is accurate, and you pursue your business development with clear goals and strategies, you won't have unnecessary overhead threatening to sink your operation. You must keep up with business paperwork even if it means working weekends and the days you'd rather be fishing. You need to know your financial status, tax liabilities, cash, and inspectors' field production. Let the good news come in small doses and don't be shocked when course corrections are called for.

If you go to full office automation, expect to spend a substantial amount for the initial hardware. You can be overwhelmed by available hardware and software decisions, so it is a good idea to wait until you've been in business half a year or more and can tell what you might need from your automated equipment. Once you know what information you need to process, you can work into questions about available software. Finally, get the hardware to fit your specific needs.

As the owner of a small business you'll have a hands-on approach to every phase of operation. But as your business grows, it is important to gracefully give over responsibilities to people you've hired. Your strongest skills should be in management: treating people well, motivating and inspiring employees, handling the company's image, and projecting the image you have created for your extended enterprise.

9. PROFESSIONAL REPORTING PRACTICES

The Document You Live By

Even a flawless inspection with a carelessly written or phrased report could open the door to your worst business nightmare and leave you staring at a lawsuit. On the other hand, careful phrasing can sometimes make up for practical oversights. International diplomatic language covers a wide range of different activities for nations and their official representatives. As representative of your business, you'd do well to model diplomatic behavior.

From your first telephone contact to the handshake and parting comment exchanged with your fee, words and images are your stock in trade. To be "perfectly frank" is a mistake. In successful business communications, perfect frankness is conveyed in meaningful silences as much as in masterful words. Business communication is never, never an opportunity to unburden your innermost feelings.

While it may not be always easy, it is always better to view your client interaction as a professional communication with at least three distinct levels.

First, there's the level of personal bonding and communication. At this stage, you build the customer's confidence and trust in you as an individual.

Second, there's the level of trust in expertise, where you deliver your service. Here you use the trust you have created to deliver your technical input. You are translating new concepts into layman's language. It must be clear. And if not simple, it must at least make a complicated subject seem understandable.

Third, there's the level of communication that carries over to the business community, where your customers convey their decisions based on their trust in you. It is here your reputation is built or tattered. What the customer concludes about your services, after reflecting on everything, will be crystallized into a few words. This after-image will be the source of future referrals or rejections.

When you do find yourself in hot water with a client, it's usually because a break on one level of interaction has spilled over into another, causing it to weaken -- maybe causing total melt-down.

You want to avoid this at all costs. It's better to give up your fee than to put a strain on this personal and professional bond. When you give up your fee you lose a couple hours of your time. When you are adamant to prove your rightness you may collect that two hundred dollars or so, but it will cost you far more in bad publicity when the loser talks to friends and acquaintances about you. Possibly a spiteful repair bill totaling hundreds or even thousands of dollars.

If you're dealing with an honest and well-intentioned individual, you can give up one skirmish and win a life-long ally. The quality of any communication can be judged only by its effect, so you might as well defer to the customer. The effect will be to defuse resentment built on your opposing position, and demonstrate to the individual that you are a reasonable and good-natured person. A future referral is the likely outcome.

Everyone encounters distant and difficult customers who seem ready to pounce and prey on you. When you realize this, you should have a red flag reaction: stop what you're doing, politely suggest another avenue for them, and leave the relationship.

That's not comfortable, and the decision means you break down the relationship, which could impact on your professional reputation. But in the long run, the business community knows what is going on. Your performance over time will tell its own story, and the predatory client that you handled skillfully may enhance your reputation for being aware and insightful.

This doesn't relieve you of the responsibility for every word, phrase, and attitude you employ. In this chapter we'll examine some inspection letter reports, as well as the HomeTech Building Analysis Report (BAR) form, which can be your most important inspection tool. Used successfully thousands of times in all parts of the country, it is well accepted by clients, lenders, and lawyers.

Comparison of Report Types

The first rule of inspection reporting, no matter what format you use, is that your report must be completed **before** you go on to the next inspection. Otherwise, your accuracy will suffer and you will miss out on one of the greatest benefits of the business: when you're done for the day, you're done. You don't need to sit up nights preparing inspection reports, and it's a bad idea to do them that way.

For many years, the only type of report that HomeTech used was the narrative report: The inspector takes notes during the inspection, then gives a verbal report to the customer and says the typed report will be ready in a day or two. The inspector dictates the report on tape while driving to the next inspection, and at the end of the day delivers cassettes to the office for typing and mailing.

Over the years, problems with this system emerged. Some inspectors simply can't drive and dictate at the same time -- they're a hazard on the road to themselves and others. Some inspectors never learn the standard phrasing that makes it possible to dictate a report quickly and efficiently. And if the cassettes are not turned in promptly, the report delivery system breaks down.

The first solution tried was a one-page summary report, delivered to the purchasers at the inspection. Its purpose was to make up for delay in receiving the complete narrative report. It worked pretty well, but its worst fault showed up when only one member of a couple was present at the inspection. The other buyer would say, "You want me to make a decision on the most important purchase of our lives on the basis of a one-page summary report? No way! I want to see the entire report before we release the contingency."

The HomeTech Building Analysis Report (BAR) was the eventual result. It is an efficient, but detailed, checklist form that is completed and delivered at the inspection, with an automatic copy for the office files. Probably 75% of all inspections done in the country today now use some form of on-site report.

ADVANTAGES AND DISADVANTAGES

Narrative. It is customized for each purchaser, it looks professional, and is well received by customers. The inspector doesn't need to have legible handwriting, because the finished report is typed or printed out.

However, it is not possible for the inspector to review the finished report because of time constraints, so there is some increased chance of error. Liability risk is increased because all disclaimers must be remembered and included in each individual report. It is very difficult to guarantee prompt delivery, even in these days of fax machines -- and the delay in releasing a contingency is a major reason real estate agents and purchasers do not like narrative reports. The administrative cost of a narrative report is high: $15 to $25 compared with $3 to $5 for an on-site report.

Important requirements for successful use of the narrative report include completing the dictation immediately after the inspection, using standard phraseology for efficiency and for your own protection, and having the staff and equipment to produce the final report quickly and professionally.

On-Site. The checklist format ensures that a systematic and complete inspection will be done. Since the report is completed, delivered and paid for on the spot, there is no further paperwork -- and almost no need for secretarial support. Important disclaimers are automatically a part of every report, minimizing liability exposure. Real estate agents love it, and will ask for it if you offer a choice.

However, legible writing is absolutely necessary, which is a real problem for some people. You spend more time at the inspection itself, but less time doing things like deliver reports to the office.

It should not take you more than 15 or 20 minutes with the client after the inspection to go through and deliver your on-site report. And the inspection itself should not take more than about 2 hours. There was a legendary inspector who took five hours to do an inspection, then another hour and a half talking to the purchasers. Everyone was angry and exhausted by this process, which was almost literally "overkill." Don't let that happen to you!

Pointers for using an on-site report include the following:

✓ Review the checklist between your first and second trips through the house, to be sure you pick up anything that was missed

✓ Fill out the administrative information before going into the house, complete report sections as you go, and leave only the summary and list of expenses to be completed at the end of the inspection.

√ Customize every section of the report with specific notes in addition to the checklist

√ Include all items near the end of their useful life in the summary list of budgeted expenses

Narrative Reports

Since you are called upon to do several different kinds of inspections in the course of a week's work, it is important to be prepared.

By far the most common inspection is for a single-family detached house that is about to change hands. This usually involves a somewhat anxious if not apprehensive buyer -- your client. It may also involve a slightly secretive seller, and an agent who is as protective of the contract offer as if it were a golden egg.

You need to provide information that is understandable, clear, professional, and the quicker the better. Nearly all the time, the HomeTech BAR is precisely what you need. It has the professional quality of a nationally standardized report, which adds to your professional image. It is fast, because you can deliver it on-site. It is less expensive than any other kind of report you can give. You walk out with a good copy and the comfort of immediate cash flow. No bills to bother with.

There are instances, however, when the standardized report form doesn't quite fit. You may be called on to answer specific questions about structural conditions or the potential for renovation or remodeling. You need to provide more detail than the form permits. You'll make notes and type up a report in letter form. Old house, multi-family, and other specific situations may also be appropriate for inspection letters.

You may be called to inspect new home construction before the contractor's final payment is made. This is a punch list inspection, and you could use a form or a letter. You will be pointing out things which must be completed before the builder gets paid, so there is a good chance you'll be called back for a re-inspection, which will require yet another check-off and a letter verifying that items mentioned earlier were addressed.

You may be asked to provide specific wording to satisfy a lender between the time of your structural report and the loan closing. When you recommend an exterminator's inspection and report, they may indicate evidence of present or past infestation of wood-destroying insects. Then the bank will get nervous and want you to write a letter restating your findings regarding structural condition.

This kind of nervousness is contagious. If you call the exterminator and discuss their observations, chances are you can give the bank the desired wording without revisiting the site. If so, it's a judgement call whether or not you charge a fee. Most inspectors don't, especially if they are doing repeat business with the bankers and agents involved. However, if you must return to the property and crawl around in a dusty, insecticide-laced crawl space, decide on your fee when you emerge into air fit to breathe. It's always appropriate to say that a minimum fee will be charged if you must return to the site.

The reinspection trip may be just as inconvenient for you as the original, but you'll be more apt to have a happy customer if you charge no more than 1/2 to 2/3 the original fee. There is, after all, less work required to check items you identified and described earlier.

In a punch list or new construction report, you can use a combination of form report and follow-up letter.

Punch List Inspection

The following is a typical punch list inspection letter.

Dear Mr. & Mrs. _____:

Here are the results of our inspection conducted on the property at _____. This inspection was conducted using standard, visual techniques. There are no warranties of any type expressed or implied.

BASEMENT HALLWAY AREA:

1. Install door stop on door to storage area under foyer.
2. Caulk, putty, paint sheetrock -- several locations in this area.
3. Touch up sheetrock around the smoke detector.
4. Remove paint and clean up carpet. Clean up all stains on the carpet, especially at the entrance to the bathroom and utility area.

FRONT RIGHT ROOM -- BASEMENT:

1. Install door stop on entry door.
2. Repaint entrance door.
3. Repaint bifold closet doors. Some areas have no paint on them.
4. Remove paint and foreign matter, clean carpet.
5. Touch up sheetrock, paint, caulk and putty throughout. Paint is quite thin in many areas.
6. Repair several large dents in woodwork around the twin entrance windows and other portions of the woodwork throughout the room. Putty and repaint.

General note: Free up all windows in the building and thoroughly clean and vacuum all HVAC ducting.

FURNACE ROOM AREA:

1. Secure laundry tub to wall so plumbing connections will not rupture.
2. Install finish plate to laundry tray -- left side of laundry tub.
3. Clean or replace return air filter.

BASEMENT HALL BATHROOM;

1. Touch up paint, sheetrock, caulk and putty -- several locations throughout the area.
2. Remove paint and other debris from vanity base and vanity top.
3. Install mirror over vanity.
4. Clean ceramic tile at shower base.
5. Install door stop on entrance door.

FAMILY ROOM AREA:

1. Touch up paint, sheetrock, caulk and putty -- several locations throughout the area.
2. Remove paint, sheetrock mud, and other foreign material from brick fireplace profile. Caulk trim material to masonry.
3. Window nearest electrical distribution panel is not fitted properly and should be repaired.

Note: Final electrical, plumbing, and mechanical inspections are scheduled within the next few days. Before a certificate of occupancy can be issued, final building and public utilities inspections are required. The certificate of occupancy must be issued before utilities can be turned on and the building occupied.

EXTERIOR REAR:

1. Repair light fixture at sliding glass doors.
2. Install duct seal around penetration of gas line through brick veneer.
3. Level up air conditioning compressor.
4. Touch up caulking and paint on entire exterior of building.
5. Remove foreign matter and clean aluminum siding. There is a significant spattering of mortar on the cantilever window at the kitchen.
6. Repair dent to cowl on exhaust vent from kitchen range.
7. Straighten gutter on the side of sliding glass doors.

EXTERIOR FRONT:

1. Touch up trim, paint, and caulking -- several locations throughout exterior.
2. Remove all foreign matter from exterior brickwork. Concrete spills appear concentrated on the brick face in the base of the stairs and front concrete porch.
3. Point up mortar -- several places on front, esp. twin window, right hand side.
4. Caulk front door threshold with a silicone-based material
5. Install flashing over front door feature trim.
6. Repaint front entrance door.
7. Provide weep hole openings in all storm windows.
8. Install splashblock -- front left hand gutter downspout.

FOYER:

1. Repaint threshold on inside and adjust weatherstripping so daylight is not visible.
2. Remove hinge stop and repair dents in trim, install base type door stop.
3. Touch up sheetrock, caulk, paint and putty -- several locations throughout.
4. Add one additional coat of paint all interior doors. Coverage is thin and missing in some spots.
5. Remove debris and clean entrance door.

FOYER POWDER ROOM:

1. Touch up sheetrock, caulk, paint and putty -- several places throughout.
2. Install and caulk side splash.
3. Remove paint and clean vanity top and base.

LIVING ROOM AREA:

1. Touch up sheetrock, caulk, paint and putty -- several locations throughout.
2. Realign and true bay window operable casements, fix handles.
3. Install shoe mold at base, left and right of bay window area.
4. Remove paint and debris, and clean the area thoroughly.
5. Bulge in carpet at wall between foyer and living room. Note this to be corrected or adjusted.

DINING ROOM:

1. Touch up sheetrock, paint, caulk and putty -- several locations throughout.
2. Replace dented thermostat cover.

KITCHEN:

1. Touch up sheetrock, caulk, paint and putty -- several places throughout.
2. Rehang or adjust cabinet doors to fit plumb at top and bottom.
3. Remove paint and other debris and clean up area thoroughly.
4. Replace stile on triple window -- right side.
5. Repair chips in kitchen counter top -- left side of range and work area.
6. Repair and repaint several dings in cabinet area finish.

STAIRWAY AND UPSTAIRS HALL:

1. Touch up paint, sheetrock, caulk and putty -- several locations.
2. Remove paint and debris from carpet, and thoroughly clean area.
3. Secure carpet to stairway throughout run of stairs.
4. Install door stop on upstairs hall closet door.
5. Repair large ding in casing at closet door.

FRONT LEFT BEDROOM:

1.   Touch up paint, sheetrock, caulk and putty -- several areas throughout.
2.   Remove debris and thoroughly clean carpet.
3.   Repair knot-hole in left hand bifold closet door.

FRONT RIGHT BEDROOM:

1.   Touch up paint, sheetrock, caulk and putty -- several areas throughout.
2.   Remove debris and thoroughly clean carpet.
3.   Repair damaged blind, right window.

HALL BATHROOM:

1.   Touch up paint, sheetrock, caulk and putty -- several areas throughout.
2.   Adjust all doors for proper fit, closure, locking, and operation.
3.   Remove paint and clean vanity top and base.
4.   Repair dings in vanity base.

MASTER BEDROOM:

1.   Touch up paint, sheetrock, caulk and putty -- several areas throughout.
2.   Remove debris and thoroughly clean carpet.
3.   Remove paint and clean vanity top and base.
4.   Install door stop on master bedroom closet.
5.   Remove any nails from stiles on window sashes throughout the building.

ATTIC AREA:

1.   Install insulation on back of access door.
2.   Install thermostatic vent fan at gable end vent.
3.   Clear blown insulation from soffit vent areas and insert permanent baffles for assured air flow.

This concludes the punchlist report. While we believe this is basically a well built standard structure with no major problems, general attention to cosmetic details is lacking. This is a production house by a production builder, and the general inattention to cosmetic details is typical. Items noted above should be corrected before settlement.

We would also remind you that the builder is required by law to warrant the house against all defects in structure and equipment for one year.

Enclosed is a copy of our Real Estate Data Sheet to help you anticipate additional expenses.

We believe general maintenance for this building should run in the $200-$500 range for the first 5-8 years, then possibly $500 to $700 per year.

Sincerely,

PUNCH LIST ANALYSIS

In reviewing this letter report, you will find several stock observations that are repeated. While these do not specifically locate a problem ("piles of sawdust on rug six feet from left side of fireplace" or "dent in sheetrock 18 inches above heat register on west dining room wall"), they do alert the client and the builder's punch list crew to the condition.

The letter functions as a checklist for the builder and the bank's inspector, or your own reinspection. It tells the builder these deficiencies must be corrected before release of the final hold-back funds will be authorized. This is the time when the individual buyer is in a position to get immediate attention from a production builder. Money talks.

Even though a punch list takes more time than a pre-purchase inspection, you know that the little nits you pick have a high probability of improving the living conditions and attitude of your client. This sort of job could be one of your best public relations moves.

Stock Phrases. You'll note that this inspection is broken down room-by-room, the method that is used nearly all the time for punch lists. In so doing, it repeats certain observations as they occur, which may be many times.

You could cover these observations by making a generalization, for example:

General Condition Observations

1. Touch up paint, sheetrock, caulk and putty -- several locations.

2. Remove paint and debris from carpet, and thoroughly clean area.

The above conditions require attention in all rooms with the exception of the master bath. Numerous sheetrock dents and nail pops are obvious. The number is 20 to 22 occurrences. The entire house is ready for final pre-settlement cleaning, which requires removal of all foreign matter from all carpets and vinyl floors, from inside heat ducts, and thorough cleaning of all manufacturer's labels and markings from all doors and window glass.

Punch list by trades. You could break down the punch list by categories of work. For example, your list could include every item the electrician will be reasonably expected to fix. Likewise with the sheetrock finishers, painters, trim carpenters, and the construction clean-up crew.

Using your knowledge of the construction process in your area, a little effort is justified to structure your punch list reports in a manner that is convenient to the industry. You might interview several bankers or builders to discover who actually does the punch list work. Sometimes builders have one crew that does the work of several different trades.

On the other hand, there will be occasion to call specific subcontractors for certain types of tasks. Electricians, plumbers, HVAC contractors, and possibly the gas company could be required for safety reasons to do any work on their systems or equipment.

It is quite likely that the contractor will have one or two specialized crews to do most of the punch list work. One would be the finish crew. They might handle minor sheetrock dents, paint touch-up, caulking, resetting cabinet doors, installing handrails and trim molding, door stops and hardware, screens, light fixture globes, repairing blemishes in shingles, and fixing downspouts and splash diffusers.

Another would be a subcontractor clean-up crew. They come in last, after all repairs are made, and when they leave the house is ready for your final re-inspection or for the mortgage lender's walk-through after all final permits and authorizations have been made.

By knowing which workers do the tasks described, you can write clear paragraphs in your report that communicate directly to the people responsible for assigning the tasks, and also tell the workers what is required to make the repairs.

Report for a Multi-Unit Building

The following narrative report comes from the HomeTech files. Client names, the property address, and the date of the report are obscured for reasons of anonymity. This report takes into account the special circumstances of the sale: clients are considering a business investment, are sophisticated in their financial approach, and are in a world apart from the first-time home buyer.

Dear Mr. and Mrs. _____:

Here is the inspection report on the five unit apartment building at _____, in Washington, DC. The following is the report on the present condition of the structural and major utility installations in the condominium conversion.

General: The building is a solid masonry structure which we believe to be 70-75 years old. Approximately four to five years ago, the building underwent a complete renovation including replacement of all major components except the roof.

Structural: Structurally, the building appears to be in very sound condition. It is solid masonry construction with wood floor framing. There are no signs of major defects due to settlement or wood boring insects. We do, however, recommend that a termite proofing warranty be maintained on the property continuously as part of the condominium maintenance program.

Electrical: Electricity is provided to each unit by individual meters and distribution panels that were updated and repaired at the time of renovation. The individual panels are 150 amp circuit breaker services with the exception of the one efficiency unit, which has a 100 amp service. This should be adequate for the present and future needs of the units. The lower branch wiring is copper and from all visible indications should meet District of Columbia code. Smoke detectors and strike stations for the fire alarm system have been hardwired (on separate, full time electrical power supply). Duplex outlets and switches tested at random during the course of the inspection were functioning normally. Individual units are pre-wired for telephone service.

Plumbing: Plumbing supply pipes are standard grade copper which was installed at the time the building was renovated. Water supply is through a single main meter which should be monitored by the condominium association. Supply piping and waste water lines do appear to be in good condition. Hot water is provided to all units by two 40 gallon gas fired heaters. These units were installed at renovation and should have a normal life expectancy of 8-12 years. Both units are hooked up in series and should provide ample hot water for eight to 12 persons.

Heating/Ventilation/Air Conditioning: Heat is provided to the three large upstairs units by a 2-1/2 ton heat pump system for each unit. The basement unit has a nominal 2 ton unit. These were all functioning at inspection and should have a life expectancy of 7-10 years. The efficiency apartment unit is heated by electric resistance baseboard convection heaters, which were functioning on inspection. Air distribution throughout the individual units does appear to be adequate. The ducting system is above standard quality with high and low registers for both summer cooling and winter heating.

Kitchens: Appliances in all units are electric, with range, refrigerator, dishwasher, and disposal. Exhaust fans in the kitchens are vented to the exterior. Normally, cooking and refrigeration equipment lasts 15-20 years, while dishwashers and disposals have a normal life expectancy of 5-12 years. All units are equipped with washers and dryers, and we recommend drip pans to be placed under the washing machines to prevent any seepage to the units below.

General Interior: Walls and ceilings of all units are plaster over solid masonry and drywall. Brick has been exposed and sealed with a penetrating sealer. Soundproofing insulation has been installed in walls and floors throughout the building. Crown mold is installed throughout. Windows in the basement apartment are aluminum sliders and wood double-hung units, all double glazed. Basement windows are protected with locking iron grills. On the second and third floors, wood double hung window units protected with storm sash are placed everywhere except the left front windows. We recommend adding storms to these units as well.

Flooring in the basement unit is ceramic tile and wall to wall carpeting, with refinished pine floors throughout the remainder of the building. Floors appear to be in good condition. Insulation has been installed in the third floor ceiling and roof sheathing. This has been brought up to contemporary energy efficiency standards and should make for a fuel efficient building.

General Exterior: Finish roof on the building is a standing seam metal roof approximately 10 years old. With proper maintenance this should last indefinitely. It is a quality job and has been coated with red lead. Roof at the rear of the building is also standing seam metal roof which has been tarred and painted with reflective metal paint. This roof should be checked annually for leaks, because metal roofs will last a long time if kept painted or coated, but if they rust through they cannot be satisfactorily patched and must be replaced. Also, when a metal roof has been coated with tar it is impossible to determine the condition of the metal under the tar. We recommend that a reserve fund be established in the condominium budget for replacement of this roof. New gutters have been placed at the rear of the main section and appear to be improperly aligned. Downspouts are draining into an underground storm sewer. Chimney and rear wall need pointing up. Fire escape needs a coat of paint. Off street parking spaces are provided.

Conditions: The structure of this building is very sound. Electrical, mechanical, and plumbing equipment are in good working order. The roof was in good condition and no leaks were evident at time of the inspection. Approximate age, estimated years of use remaining, and replacement cost ranges for major components are as follows:

Item	Approx. Age	Use Remaining	Replacement Range [Per Unit]
Bathroom fittings and fixtures	4 years	20 years	$1,500 /$2,000 per
Kitchen sink and fittings	4 years	16 years	$200/$300
Disposal	4 years	6 years	$100/$175
Kitchen range	4 years	16 years	$400/$550
Refrigerator	4 years	16 years	$500/800

Waste water piping system	4 years	26 years	$4,500/6,000
Water supply	4 years	26 years	$4,800/6,400
Heat pumps	4 years	6-8 years	$2,200/3,000
Domestic hot water heaters	4 years	8 years	$350/500
Roof	10 years	30 years	$6,500/8,500
Lighting fixtures	4 years	16 years	$800/1,400
Fire alarms	4 years	26 years	$1,100/$1,600

Code violations: While we do not provide Building Code inspections, there are no apparent code violations on the property and none recorded or outstanding against this property on the inspection date.

Sincerely,

REPORT ANALYSIS

Systems approach. Notice that in the report above, the systems approach was used. The inspector did not identify specific units, much less describe them in detail. This report applied to the whole building and addressed the questions which might be of interest to the new owners. In general, what overall conditions were found in the structure and equipment? What are the projections for replacement of the various systems, both in terms of cost as well as time lines? Are there outstanding problems which could create a legal hassle due to present or past code violations?

The inspector in this case had to know the clients' intended use of the structure in order to provide the most pertinent and useful information. You may note certain standard language relating to the coated metal roof, and the use of replacement cost ranges. These issues are covered in more detail elsewhere.

Lender's Wording

When you recommend the services of a licensed exterminator in your initial inspection report, you may be called for a reinspection or a letter to soothe a lending institution's closing officer. In the following letter, an insect report raised questions that could have had structural implications.

When you are asked to submit a structural letter in similar circumstances, your own warning light starts flashing. Obviously the lender is looking for an authority to assume legal liability, taking it off the back of the financial institution.

Caution on your part is completely in order. It's a judgement call whether you feel confident enough to sign the requested statement without a reinspection. The specific nature of the subsequent finding after your inspection must be fully disclosed to you before you can make a reasonable decision. Have the lender or real estate agent (whoever has requested your letter) read you exactly the wording of the termite report. You may wish to have a copy of this report for your files, which is also prudent and reasonable on your part.

Next, if possible, call the exterminator and chat with the person who made the termite report. You can talk very specifically with this individual, who shares with you a clear vision of what the structural underside of the house in question looks like. From that conversation you can probably decide whether a reinspection will be needed.

Unfortunately, time is usually a major heartburn. They don't know they're definitely going to want that in the closing package until a day or two before the scheduled closing. You don't want to obstruct the process and offend a good client who will do more business with you in the future. You don't want to make a reinspection, but if you must you definitely don't want to be under the stigma of having missed something. And you don't want to do it free of charge.

On the other hand, you're busy. You've already done your work. This is an inconvenience. Even worse, it creates self-doubt. Is there something significant you might have missed? Did you fail to belly over into the far corner of the space? Is there a liability dragon living there?

Your first instinct is to waffle-word:

XYZ Federal Credit Union

Re: D-------------- Property
     Address
     City, State, Zip

To whom it may concern:

We made a thorough inspection of the above captioned property on May 8, this year and discovered no substantive structural damage. There are cosmetic repairs needed in several instances but the structural frame of the house is essentially sound.

The house is structurally well built and should require only moderate maintenance over the long term.

If we may be of any further assistance in the matter, please contact us at your convenience.

Sincerely,

The terminology above was not completely reassuring to the lender. In fact, it probably did not make a significant difference in the liability exposure of the inspection company. By specific request on the telephone, the loan officer elicited the following wording:

"We made a thorough inspection of the above captioned property on May 8, this year and discovered no structural damage at this time. Items noted on the termite report dated May 22 are cosmetic in nature, and water damage is due to weather conditions."

The original inspection report was made using the standard HomeTech Building Analysis Report (BAR) form. It could have been anticipated that the insect report would contain observations about past infestations and wood damage. However, the reaction of the lending officer and the specific desired wording were not standard on an industry-wide basis. When the release required must be worded specifically, you have little choice but to comply, provided the conditions described are accurate to the best of your knowledge.

You're often staring down the long barrel of a legal liability. Your service rides a thin line between giving all you can to keep the public satisfied, and sticking your neck out.

Reinspection After Repairs

Sometimes the situation takes on a tone of enforcement. In the letter below, a real estate settlement was being delayed by the buyer's attorney pending completion of a list of items pointed out on the initial inspection, which was made on the standard HomeTech BAR form. Tempers were strained because the original inspection had been called for at the last moment from the attorney's closing table. Because of a sense of mistrust, the attorney refused to go to settlement until the repairs had been made and certified.

The certification letter required another inspection beneath an old house, inspection of specific items, and the following letter. All was under heightened time pressure, with the result that a reinspection fee was justified.

RE: R---- Residence
Address

Dear A----:

At your request, I made a reinspection this morning of the fireplace and crawl space area at the R---- residence noted above. Items noted in my inspection report to Mrs. E---- E------ on May 30 have been substantially addressed and I found no further immediate concerns.

More specifically, abandoned wiring and old wrapped steam piping have been removed. The wood debris has been removed and crawl space earth raked clean. Insulation batts which had fallen from the floor joist area have been reinstalled. Damaged sills have been reinforced and new piers installed for support. A new sill has been installed to provide support for the fireplace hearth and brick face, and the cracks noted earlier inside the firebox have been repaired with fireproof mortar. This reinspection was made at approximately 9:30 a.m. this date.

Please call me anytime I can be of further assistance.

Sincerely,

The Building Analysis Report (BAR) Form

The HomeTech system of home inspection is being updated continually, based on use in the national marketplace, by HomeTech staff inspectors, and review by outside hired consultants. The BAR forms make it possible for you as an inspector to complete job after job entirely on-site, collecting your fees as you go. By taking careful notes on-site and clipping these to your office copy of the carbonless form, you maintain an accurate file which will give you the correct information if callbacks or follow-up letters are required.

To truly understand the nature of this business, you must know houses, multi-family buildings, people, report writing, and pets. While you're on your own in certain spheres, the HomeTech BAR form has been designed to assist you continually throughout your inspection process. Let's go through it from the beginning.

Cover Sheet

The cover sheet of your BAR contains more than client information, property address, and payment record. The inspection is described as, "visual inspection of readily accessible areas of this building," which clearly indicates that you are not to be held to account for items that can't be seen readily.

It means you don't move furniture to check outlet boxes, you are not required to climb into attic crawl spaces through ceiling scuttles, and you are not obligated to skinny under the house if the only access is a hole where a couple of concrete blocks have been left out and a board placed alongside the foundation for cover, and clearance below the joists is less than 16 inches.

You agree to do your best to conduct a reasonable visual inspection. This means you look at a lot and you deduce a lot. You are allowed to make generalizations based on your knowledge of construction and the general practices of the inspection industry. In the case of an inaccessible attic scuttle through a ceiling 12 feet off the floor, you must make a careful examination of the contributing conditions which may lead you to reasonably conclude things about the attic. You don't have to climb into the attic, and your agreement is carefully worded to limit your responsibility in such a case.

Now, the fact that you may not be required to take personal risks and get filthy wriggling into inaccessible crawl spaces does not prevent your doing so. Personal dedication to the profession may lead you to go beyond the specific limits occasionally, especially if a bouncy floor makes you suspect a sinister condition in the crawl space. The extra effort almost always results in a sense of relief at having seen something hidden and crucial; however, the cover sheet is written to try to contain your responsibilities and liabilities.

Pre-inspection Agreement

The purpose of this agreement is to put the relationship in understandable terms. For your sake, it limits your liability. For your client's, it explains what your inspection covers, who you are working for, who owns the report and has any right to use it, and what categories of conditions are specifically excluded from the report.

With today's environmental consciousness, one paragraph in this agreement is particularly important. It gets you off the hook in the event of needed specialty inspections, and excludes your work from identifying common but tricky hazards.

"The inspection and report do not address and are not intended to address the possible presence of or danger from any potentially harmful substances and environmental hazards including but not limited to radon gas, lead paint, asbestos, urea formaldehyde, toxic or flammable chemicals and water and airborne hazards. Also excluded are inspections of and report on swimming pools, wells, septic systems, security systems, central vacuum systems, water softeners, sprinkler systems, fire and safety equipment and the presence or absence of rodents, termites and other insects."

The language is legalese. It not only limits your customer's expectations from your inspection, it helps create a market for specialty inspection services. Most of these are done by other companies. Some are add-on services that may be done by your own company, such as radon and asbestos testing. These additional services can double your total income from services on a house. But if you did not specifically exclude these hazards by agreement from your basic general inspection, sooner or later the oversight would come back to haunt you.

Potentially the most important terminology in your pre-inspection agreement is included in the last paragraph. The intention is clear to limit your liability from oversight or inspection errors, as well as the untimely or premature failure of any system or equipment. The paragraph becomes a bargaining chip for you in unfortunate circumstances.

"The parties agree that the COMPANY, and its employees and agents, assume no liability for the cost of repairing or replacing any unreported defects or deficiencies, either current or arising in the future, or for any property damage, consequential damage or bodily injury of any nature. THE INSPECTION AND REPORT ARE NOT INTENDED OR TO BE USED AS A GUARANTEE OR WARRANTY, EXPRESSED OR IMPLIED, REGARDING THE ADEQUACY, PERFORMANCE, OR CONDITION OF ANY INSPECTED STRUCTURE, ITEM OR SYSTEM."

Laws differ from state to state. Some courts have held, and many lawyers will argue, that limitation of liability and exculpatory clauses by professionals should not be valid and enforceable. HomeTech believes this position is wrong, but the law remains unsettled. In any case, business reality tells you that to go to court in a defensive position means you lose.

When you and your customer have signed the Pre-Inspection Agreement, you're ready to begin the reporting process. At the bottom of the cover page is a record of payment which you can take care of when you're sitting at the table with your customer and beginning to explain the filled-in report. The form provides a protocol, and takes away any awkwardness in timing or asking to be paid for services.

Remarks

Following your cover sheet, the report contains seven pages of clearly written descriptions of systems and conditions, including definitions of terms used. The general remarks on Page 2 are well worth your attention. You are communicating those terms with every report you deliver, and if you don't have the material memorized, you'd better at least know what you are delivering.

The terms repeat and reinforce language introduced in the Pre-inspection Agreement, emphasizing aspects to ensure they won't come back on you. These terms are the result of years of practical experience and tens of thousands of reporting transactions. Their importance to you cannot be overstated.

THROUGHOUT THIS REPORT WHERE THE AGE OF APPLIANCES, ROOFS, ETC., IS STATED, THE AGE SHOWN IS APPROXIMATE. IT IS NOT POSSIBLE TO BE EXACT, BUT AN EFFORT IS MADE TO BE AS ACCURATE AS POSSIBLE BASED ON THE VISIBLE EVIDENCE.

WHEN ANY ITEM IN THE REPORT IS REPORTED TO BE "SATISFACTORY," THE MEANING IS THAT IT SHOULD GIVE GENERALLY SATISFACTORY SERVICE WITHIN THE LIMITS OF ITS AGE AND ANY DEFECTS OR POTENTIAL PROBLEMS NOTED DURING THE INSPECTION.

These paragraphs are presented capitalized for emphasis. The special definition of the term "satisfactory" is paired with a further explanation later on about roof covering. That explanation reads as follows:

"Satisfactory" Roof Covering

When the report indicates that a roof is "satisfactory," that means it is satisfactory for its age and general usefulness. A roof which is stated to be satisfactory may show evidence of past or present leaks or may soon develop leaks. However, such a roof can be repaired and give generally satisfactory service within the limits of its age.

If you filed an inspection report without the precise definitions above, you might leave the customer believing that you could be called to account. If you said a roof was likely to last five more years and it started leaking the next fall, you could get an angry accusation. A roof repair out-of-pocket would be a realistic prospect for you to face. While you might avoid the expense, the customers would probably believe they had been misled.

But with the terms so clearly defined and repeated, the customer has to admit that he/she was informed about the limits of the inspection. Those words could well be worth hundreds or even thousands of dollars to you.

Your Pre-Inspection Agreement separates you from liability for environmental hazards. If you should overlook asbestos insulation on the steam pipes of an attic, you're covered. The wording in your Remarks section is again quite clear and specific:

Asbestos and Other Hazards

Asbestos fiber in some form is present in many homes, but it is often not visible or cannot be identified without testing.

If there is reason to suspect that asbestos fiber may be present and if it is of particular concern, a sample of the material in question may be removed and examined in a testing laboratory. However, detecting or inspecting for the presence of absence of asbestos is not a part of our inspection.

Also excluded from this inspection and report are the possible presence of or danger from radon gas, lead paint, urea formaldehyde, toxic or flammable chemicals and all other similar or potentially harmful substances and environmental hazards.

Correct use of the BAR form is not intended to be blanket coverage for sloppy inspection work. It is, however, the best protection you can employ to allow you to go about your work diligently without the nagging fear of liability exposure. Even a flawless inspection must be carefully reported, and careful phrasing can sometimes make up for practical oversights.

Making Out the Report

After some use in the field you'll be comfortable with the reporting method outlined in the HomeTech Building Analysis Report. In general, you'll be trained by using the document to look for specific items while going through your inspection. Although the house may not seem conducive to a systems approach, your eye and brain will quickly adapt, and your field notes will be easily categorized. While on site, you'll be thinking in these terms and recognize how logically the categories fit together.

STRUCTURAL AND BASEMENT

Most structural problems will be accessible through the basement and seen from this area. You will pick up clues while viewing the house from the exterior -- sagging ridges, etc. -- which your basement visit will help you follow through. In houses with crawl spaces, basement categories do not apply and the order of the inspection is often reversed: You come out of the crawl space dirty after having done the rest of the inspection.

However, after using the BAR form repeatedly, you learn to think in the order of categories presented. The logic justifies itself.

The category "Structure" can cover a vast collection of specific details in a simple check-box: "No major structural defects noted -- in normal condition for its age." This seldom meets your wish to provide positive observations, so there is space for your written comments. Some typical ones:

Historic home. "This historic house was restored 11 years ago. Extensive upgrading and addition of wing and garage was done at that time. New work appears to have been completed in a workmanlike manner. Restoration work is above average."

Owner-built home. "Structural frame is above average; however, water damage has occurred to siding due to improper maintenance (painting), improper finish joints of wood to masonry, and improper backfilling of earth to wood at front planter. Repairs could be extensive."

Condominium on ocean front. "Building is well-maintained, units are built to national code. Some decomposition at sliding glass door and on deck due to salt/moisture. General building maintenance appears good. Repair/maintenance program is presently underway."

Interestingly, in a vacation cottage you might encounter a list of several structural items and still check the box, "No major structural defects noted . . ." The question you're dealing with is how to point out the defects and avoid stampeding your customer. In such a case, you might use the space below the word "Structure" to put a number of the items you have identified. For example:

1. Roof rafters decomposed at end against chimney due to continuing water leak.

2. Soffits rotted, need replacement on North, East, South sides of house.

3. Porch beams dry rotted, need rebuilding

4. Front door frame metal rusted through

5. Fascia boards rotted (aluminum clad over)

In this case the prospective buyers had let the inspector know their intent with a 30-year-old, flat roofed, concrete block beach cottage. Since major renovations were planned and the owners would do the work themselves on a part-time basis, the inspection report presented a general scope of work.

HEATING AND COOLING

This section is well detailed and should require very little elaboration. Space is provided for explanations as necessary.

PLUMBING AND BATHROOM

The category is self explanatory. Using the boilerplate remarks provided, you can checklist through the entire system. Of course your inspection must be thorough, and there are spaces provided for your comments at each category. It is a good idea to repeat boilerplate wording in any notes. This ties your report to the explanations and protections, and directs the client to read the Remarks section.

ELECTRICAL AND KITCHEN

The electrical section trains you to look for certain categories of items, e.g., copper branch wiring, smoke detectors, ground fault interrupters, the size of the service, breakers, etc. In nearly all your inspections these checklist items will make you see everything that needs to be covered. Occasionally you may find outdated knob and tube wiring, undersized entry service from the transformer, an inadequate ground, etc. While these are unusual, they must be specifically described in the note spaces on your report. Improper grounding of an electrical system is particularly important because it creates a potential safety hazard, which you must be prepared to describe.

INTERIOR AND ATTIC

The checklist portion is complete and useful. However, the "Windows" category fits into the Exterior section as well as Interior. Testing windows for operability must be done from the inside, but you may well discover wood rot, weak and missing sash weight cords, glazing in need of repointing, improperly vented or drained storm sash, and other items which you should note briefly in both sections of your report.

Attic insulation and ventilation are in one category. Many inspectors add comments and findings on floor and wall insulation in this category. Floor insulation is also checked in the Crawl Space, but a concern as important as energy efficiency in a home deserves repetition in your report.

ROOFING AND EXTERIOR

The Remarks section provides clear explanation of types and conditions of roofing. Your written notes on shingle condition, along with reasons why the condition exists, help the customer understand repairs and what precautions to take in the future.

Spaces are provided for notes on Doors and Windows. Here again you can use repetition to drive home a point. (As you may have noticed throughout this study manual, repetition is a major learning tool.)

GROUNDS

Attention to water drainage is such a critical issue that it is given more space than structure in the BAR. Your comments are both informative and motivational, as many homeowners are willing to undertake landscaping projects when they buy or remodel a residence. If customers are feeling overwhelmed by construction technicalities in your report, at least here, on the grading question, they can feel empowered to take action. Prudent comments are well accepted here. They have a good chance of improving your popularity with the customer.

The Most Important Part

The Summary page is the most important part of your report. Many customers feel more at ease if you show them this part first, then work in the details that support your conclusions. Everybody is looking for a way to describe the condition of the structure in a few words. When you can reduce it to a range of two cost figures and a one- or two-sentence remark, you've given them the foundation for their bargaining or financial strategy.

Certainly your analysis has to proceed from the specific to the general if it is to have meaning for you. But for the customer, the opposite is often most manageable.

You're dealing with short-range maintenance and repair numbers as well as medium-range expense probabilities. In many cases items will fit into both categories, as the repairs needed immediately will also be items needing replacement or repair within the next five years. Many inspectors, in this circumstance, place an asterisk beside the items on the medium-range list that are included in the "first year" list, with a note of explanation.

"Items not operating or with major deficiencies" might include the following comments:

Garage door broken
Siding repair two areas
Smoke detectors
Ground fault interrupters
Screen room collapsing roof panels
Soffit vents and attic vent fans needed (not installed)

Following down the page, the next list is, "List of some important items requiring possible repair or replacement within next five years:"

Paint house	$900 - 1,500
Re-roof	1,400 - 1,800
* Repair siding	400 - 1,600

* Replace garage door	400 -	600
* Replace screen room roof	150 -	250
* Clean and seal deck	100 -	150
Finish garage	300 -	500
* Soffit and attic ventilation	300 -	500

Probable expenses within 5 years	$3,950 - $6,900

Items with asterisk * need immediate attention.

That tally is carried down to the category, "Miscellaneous and minor repairs and expenses during the first year of occupancy are estimated to be:" $1,950 to $3,700. This number is derived by adding the indicated items plus a general maintenance factor of $600 per year, which is considered realistic for the house in question.

Be careful about listing more items than necessary, especially for the first year. It is easy for a purchaser to be overwhelmed by the list of expenses and decide not to buy the house. Make it very clear how items can be spaced out over months or even years.

General maintenance categories would include such things as pruning trees and shrubbery, cleaning mildew off siding, power washing, painting windows, replacing several screens and light fixtures, service calls on air conditioning condenser and water softener, etc.

The last item is the inspector's remarks. This should be informative, upbeat, and optimistic. The following remarks would apply to the house in question:

"House is structurally well built but needs repairs due to delayed maintenance and oversight at time of original construction (attic ventilation). After repairs to upgrade these deficiencies, the residence should be serviceable for an indefinite time with modest regular maintenance.

"Have exterminator inspect for insect damage."

Using the BAR With a Multi-Family Building

When you inspect an individually-owned unit in a multi-family building or cluster, your report cannot be confined to what you discover within the four walls.

The client is buying into more than a single-ownership house. If your inspection is to be truly useful, you must assess the situation as an inspector/investigator. You'll have to talk with real estate agents, building maintenance people, and possibly independent maintenance services in the area.

Your findings can still be covered by the HomeTech BAR. However, the Summary page will include other information. For example, here are Summary remarks from several different condo/town house inspections:

1. This is a modular unit assembled into a 16-unit building.

Management recommends replacing old compressors with new Arco-Air unit.

Important safety factor to get GFCI working to prevent electric shock injury.

Unit is well built, needs maintenance cycle replacements.

2. This is a three-level town house condo being bought new.

The property is well-sited, landscaped and built to present standards. Items cited for correction are normal punch-list items for a typical new unit, but should be corrected before closing.

3. This is one unit in an 8-unit building, in a five-building complex.

Unit is fairly well maintained but has repairs needed on a continuous basis. Building has original roof, needs exterior stain. In the past four years, dues plus special assessments for this unit have been under $1,500 per year. You should verify that the association has reserves of $104,000, anticipates replacing three roofs this year at $12,000 per roof; then one building per year thereafter. Sewer system repairs are underway, could be a major expense judging from similar experiences in other complexes.

4. A unit in a very substantial concrete/steel building, with association having a history of political infighting.

The unit is in almost new condition; building, equipment and grounds are well maintained. This inspection covers physical visible conditions only and does not address any issues which may or may not be current with the condo owners association.

Special Circumstances

Partial Inspections

You may occasionally have a partial inspection for a specific problem. Quite often a financial institution is involved. A question may need to be answered before the exact dollar amount of a loan or sale price can be ascertained.

You may have time to do this inspection at a reduced charge. However, you're not just "dropping by to look it over." If that were the case, the bank's loan officer would drop by on the way from lunch or another progress inspection. You're making a call in your official capacity, and if you doubt that for a moment, wait until the parties call you to ask detailed questions about your discoveries.

Making the inspection and writing the report will cost you as much as a complete inspection. If you do the work at a reduced price because you're willing to go along with the idea of less pay for less work, you're shooting yourself in the foot. As HomeTech tells inspector trainees repeatedly, "Get paid for what you know, not what you do."

It actually takes more work to do a partial inspection and write a complex letter report than it does to do a full inspection using the BAR form. You're not doing less work.

The following letters were issued to cover partial inspections of older houses which were about to be sold. The owners requested that a full inspection be avoided. The inspector was cooperative.

FLOOR SYSTEM IN AN OLD HOUSE

The inspection turned up a large rat's skeletal remains, among other things.

Dear Mr. V----------:

In your company this morning, I made a careful visual inspection of the structural floor system in the residence at 108 N. 9th Street. The purpose being to assess the extent of structural damage due to moisture and insect infestation, my inspection was limited in scope and did not include electrical, mechanical, or plumbing systems. By mutual agreement made prior to this inspection, liability for oversight or error will not be the responsibility of this inspector and no liability is assumed by the fact of having made the inspection and issuing this report.

The house was originally of superior construction for its day, having substantial brick piers with ornamental brickwork curtain walls between, and there does not appear to be any problem from settlement in the foundation. The house was constructed in two phases, an east wing consisting of kitchen, pantry, dining room and porch having been added subsequent to the original construction. The structure is balloon-framed on heart pine sills and has a floor system of 2x10 and 2x12 rough cut pine joists.

In the recent past, numerous stacked concrete block piers were installed at mid-span between the sills, supporting minor sills of 4x6 pine. This was done as a stiffener for the floors and appears to have made a substantial improvement; however, there is structural weakening due to insect infestation and several joists (six to 10) are slightly weakened as an effect of beetle action. This weakening is apparent only by testing the individual joists and does not result in a "springy" floor, due to the mid-span sills. We are not qualified to state whether there is ongoing insect activity, and offer no opinion on the matter.

The crawl space ground beneath the house shows evidence of water sweeping, and regrading at the exterior of the house should be undertaken to stop this water movement and its resulting moisture damage. The finish grade should be sloped away from the foundation wall for a distance of four feet, at a slope of 1" to the foot, and a swale or other means of water control installed. Also, gutters and downspouts, each downspout equipped with a splash block to direct water flow away from the house, should be installed.

Structural Damage: Several sills have incurred major structural damage from insects or moisture, and should be removed and replaced. In addition, a major bearing sill is cracked at mid-span along the west wall of the house and should be supported with a new pier at the break. In addition, joists under the north end of the front porch are collapsing due to wet rot and should be replaced or repaired. The most significant damage was found along the west porch and the southwest corner of the main house, and this should be addressed immediately because collapse of these sills could eventually result in major settling of the two-story house structure.

The specific damage and approximate cost of repairs are listed below:

1.   Main west wall support sill, SW corner. Damage due to insects is causing complete failure of the sill. Replacement of the sill is needed, at a cost for contractor's services of $400-800.

2.   SW corner of two-story porch, main sills at the corner are badly decayed and are crumbling. Damage due primarily to moisture. Replacement of the two sills is needed, at a cost of $800-1,000.

3.   Moisture damage throughout porch structure does not appear to threaten imminent collapse, however, additional stacked piers should be installed at a cost of $250 each. Four to six piers, $1,000-1,500.

4.   North end of porch, joists are crumbling due to longstanding moisture damage. Two or three joists should be repaired by "sistering" new pressure-treated joists to the existing ones, removal of the rotted ends of joists, and re-closure of the porch floor. Estimated cost, $150-250.

5.   Wooden entry steps at the rear porch are badly decayed and pose a danger. They should be rebuilt as soon as possible before injury results. Estimated cost, $150-250.

6.   In addition, there is no insulation in the floor system and too many interruptions on the ground surface to reasonably recommend a moisture barrier ground cover. The floor system should be insulated with R-19 batts between the joists and under-wrapped with Tyvek or other wind breaker wrap. This is a general recommendation and does not impact on the structural integrity of the house.

7.   Also, ventilation beneath the house should be increased, which may be accomplished by simply installing 1/4" mesh hardware cloth panels beneath the porches and other openings, leaving the ornamental brickwork in place. The improved ventilation of the crawl space should be undertaken along with

regrading the finish grade at the exterior of the foundation walls, and installing the Tyvek wrap (at minimum, even without insulation) on the bottom side of the floor joists.

Conclusion: The house in general is well constructed and worthy of restoration. While there are obvious areas needing attention that were outside the scope of the present inspection, the structural condition could be brought up to normal standards of safety at an estimated cost of $1350 to $1935. Additional repairs including painting inside and out, refinishing floors, new wiring, insulation, and underpinning, would likely cost in excess of $10,000 before the total restoration is in place.

Thank you for this opportunity to be of assistance.

Sincerely,

Another floor system inspection report follows. Notice the disclaimer of liability and specific language limiting the scope of the inspector's responsibility by mutual consent. This language reiterates what is in the pre-inspection agreement signed by inspector and client.

Re: Floor system inspection
138 D---- Blvd., Atlantic Beach

Dear Mr. G-------:

Today I made a first-hand inspection of the structural floor system of the above residence. At your request, I looked specifically for structural damage that might be the result of insect infestation and found none.

However, there is structural damage to the floor system apparently resulting from wet rot; and in addition bearing girders have been notched radically, apparently by a chain saw, for reasons unknown. In several instances the girders have been reduced to a cross section of less than 4 x 4 inches for a distance of some 16-24", with joists bearing over the center of the unsupported opening.

It is my recommendation that these girders be repaired by reinforcing the breaks with spiked-on 2" lumber stock each side of the notched area wherever a joist rests in the middle third of the notch, or by shoring up beneath the center of the notch with a support resting on a concrete pad on the ground.

In addition, the subfloor is rotted extensively beneath the bathroom on the western end of the house and should be replaced, along with whatever supporting structure is discovered defective when the structure is disassembled for repair. This same situation occurs beneath the kitchen sink, although the rotted area is much smaller. Also rotted and in need of replacement is the mud sill resting on the concrete foundation wall for approximately six feet in both directions from the northwest corner of the entry stoop. This is advanced rot in the sill and the wood is powdered, and damage can also be seen from the outside siding near the lower corner of the front door.

It is my opinion that the above cited repairs will solve your problems with the floor system.

Other cosmetic damage was noted but is not structural in nature and was specifically not included in this inspection. That includes siding below the southeast front window and broken wood shingles in several places around the house. I did not inspect interior condition, wiring, equipment, roof, etc., as this was excluded by your instructions to me.

The above report is made using normal visual means and close personal visual inspection. By mutual agreement, you will not hold me liable for errors of commission or omission. Thank you for this opportunity to be of service.

Sincerely,

Summary

As a home inspection enterprise, your business will rise or fall on the quality of your work and power of your personality. But the longest lasting record of your work will be the reports that remain stored in the document files of mortgage lenders, real estate attorneys, homeowners and real estate agents. By these documents you will be remembered, and by these your business could succeed or set up its own self-destruction.

Careful inspections are important, but no more important than careful reporting practices. You will be called on to educate the public with clear, understandable terminology every day you are in business. Much of the clarification you give will be spoken face-to-face. Much of it will be keyed to written phrases you must use correctly both for your own protection and for the public's understanding.

As a home inspector you are a practicing building expert. A professional dealing in concepts and information, you bear responsibility which cuts both ways. You are not only responsible for how you observe conditions in the field, and how you report them. You are called also to verify that repair conditions have been met.

You must be extremely careful in the wording of these verifications, both to meet the receiver's needs and to avoid creating a long-term repair liability or worse. If you specify that certain repairs meet structural or safety standards and there is a subsequent failure resulting in a building collapse or personal injury, disaster may strike. Worst case scenario, you could be erased as a business and stripped of your personal assets, perhaps in debt for life. Don't let it happen to you!

Without question, your best practical protection is the kind of careful state-of-the-art wording contained in the HomeTech BAR. Your report forms are a collection of powerful concepts and information. Use them to help standardize your own services, and to tap into a national standard. Financial and real estate professionals throughout the country will quickly come to appreciate the power and content of these general form reports. You will quickly come to appreciate the easy exchange and record keeping system they create.

In special circumstances, you will be called to provide partial inspections which can't be handled with the report form. Letters will be required. You can store generalized phrases in your computer and construct letters by selecting the appropriate blocks of words. Or you can write each one to meet the needs of the moment.

If you provide letters for special inspection services, be sure to include a paragraph disclaiming any responsibility for oversight and omission. The term "good faith inspection using standard visual means" can be a beneficial tool in any narrative report. But without a signed pre-inspection agreement verifying that the client understands and agrees to this provision, these words may have no legal standing.

The practice of home inspection may require more than a visual inspection of conditions, especially when houses are under construction and plans are on file, permits are to be issued after certain tasks are completed or in place which may not appear physically on the site. In multi-family structures, a visual inspection may also include specific financial data which comes from sources other than your site visit.

Sometimes your role is that of inspector/investigator. In a condominium unit, the report will include information gained by interviewing maintenance people or reading association documents. You can't provide a complete inspection service to a prospective condominium buyer without seeing more than what appears inside the walls of the unit and the common areas of the building and grounds. If you can't address the association's activities and financial standing, this should be included in your report. Be careful not to assert what someone else tells you as fact unless you can verify it. State clearly that your client should verify information reported by third parties.

Sometimes your role skirts close to the work of designers, engineers, and architects. While you are strongly inclined to give customers what they perceive is the best and most complete service possible, **play safe.** Check your local and state regulations. You don't want a county building inspector with a badge coming after you for observations you might include in your report about code compliance. And you don't want to receive a nasty letter from the Board of Architecture regarding your practices which might infringe on their territory.

Home inspection is a challenging field and a rewarding occupation. The liabilities are manageable, but you must exercise due diligence every day.

10. LEGAL ISSUES

Legal Considerations for the Home Inspector

As a home inspector you must be aware of the legal pitfalls associated with the industry. Because the business is fairly new, there are few legal precedents and a limited number of court cases to assist you in avoiding liability traps.

This chapter does not aim to give you specific legal advice, but you will learn certain strategies to avoid claims or defend yourself during the daily conduct of your business. Contents of a contract agreement, a note-taking form that will stand up in court, methodical inspection procedures, careful writing and disposing of your reports, and avoidance and defense of liability claims are discussed here.

It is essential that you know your own state laws regarding the home inspection field. At this writing, only one state, Texas, requires a license for home inspectors. There is a trend toward licensing, as more states are considering licensing and others may well do so within the next few years.

This is not necessarily something that the industry should oppose, because licensing will extend a minimum standard. On the other hand, inspectors who are doing an outstanding job in the field today could find that performance standards are lowered by a licensing exam, with an ensuing flood on the market of semi-qualified inspectors and overall erosion of the quality and value of the service.

Consumer protection laws in your state will have a bearing on how you conduct your business. Each state has an agency to deal with consumer problems and complaints.

The American Society of Home Inspectors, Inc. (ASHI) and its regional associations have formulated standards for the industry. In addition to providing an "umbrella image" that is recognized everywhere, membership in a national professional association and adherence to industry-wide standards of practice can help reduce the possibility of legal claims against your company.

Liability is, of course, a two-edged sword. It is the major negative fact about the home inspection business, but it is also the reason why the business is flourishing throughout the country. What it comes down to is this: Would you rather have no liability and no business, or a lot of business and some liability?

A largely beneficial effect of increasing liability is that marginal home inspectors are being eliminated from the business. It is no longer prudent to do the occasional home inspection as a sideline, with little or no training. The business is constantly changing, with new products and problems all the time (asbestos and FRT fire retardant treated plywood are outstanding examples), and it's just too risky for a home inspector not to keep up to date and knowledgeable about liability issues.

Legal Counsel

Number one priority is to retain an attorney. The attorney will help you to decide whether or not to incorporate, handle the legal aspects of start-up, and advise about the content of contracts and other legal documents you will need.

Incorporation: If you decide to incorporate, an attorney will organize your corporation according to state laws. Incorporation is the preferred business format for protection of an individual business owner. It can protect against loss of personal assets in the event of a claim and has other legal and financial advantages, as well as certain disadvantages, which your attorney can explain.

Legal documents: A professionally operated home inspection business will use certain documents that can minimize exposure to liability claims. The most important of these is the pre-inspection agreement. Your attorney can assist you in selecting the correct terminology for this agreement and advise you about other documents, such as wording for your written or printed reports and strategies for defending a claim.

Playing Fair

The assistance of a competent attorney is very important. You must choose someone with whom you feel comfortable and confident. You want to operate in such a way that you protect your own interests while creating a sound framework for your attorney if an action ever is brought against you.

Now it's your turn -- play fair with your attorney. If a claim should arise, be honest with your attorney. Keep good records so you can furnish all the information necessary for a proper defense.

Appellate Court Rulings

Three ground-breaking appellate court rulings exist. New rulings may be entered literally any day, but cases in New Jersey, Connecticut, and Indiana will be discussed at the end of this chapter. By then, you will be familiar with the terminology used in the rulings and their ramifications for the home inspector.

Why is Liability So High?

The sometimes stunning level of liability found in the home inspection business did not come out of nowhere. There are at least four factors that can explain why liability is so high for professional home inspectors.

Problems With Purchasers

Many purchasers of houses simply have no idea what a home inspection should include. The business is relatively new, and professional standards are still evolving today, so it's no surprise that public awareness is limited. Frequently the only reason purchasers schedule a home inspection is that someone they trust told them it was a good idea.

Even when the purchasers are present at the inspection, they are not really focusing on the inspection itself. They're worried about qualifying for a loan, wondering whether the schools are good, afraid their furniture won't fit in the rooms, and so on. And it is difficult for the home inspector to explain clearly what will and will not be done without sounding totally negative, and making the purchasers wonder if the whole thing is worthwhile. The pre-inspection agreement does its best to provide the necessary education, but even so, many of today's customers just don't understand the limitations of the inspection process.

The Dreaded Call-Back

Suppose the inspection goes beautifully, the purchasers loved the inspector and the report, they move into the house and everything is wonderful. Then some time in the first six months something goes wrong, and the homeowners call the inspector. The call isn't returned immediately, so they call an appropriate service person -- roofer, plumber, HVAC mechanic -- to look at the condition. Or maybe in the course of a regular service call, perhaps on a gas-fired furnace, a serious condition is found such as a cracked heat exchanger which requires replacing the furnace.

One of the things the homeowner says to the tradesman is, "By the way, I had this house inspected before I bought it." Unfortunately, maybe one time in a thousand the tradesman will reply, "Oh, they could not have seen this condition in a visual inspection. It is not their fault." The tradesman wants to look good, and justify charges for service work, so it's just too easy to suggest that the home inspector slipped up. From that point on the home inspector is on the defensive, and will probably be held liable for any corrective procedure.

It is fear of bad news like this that causes many inspectors to hesitate returning calls from old customers. But that is exactly the wrong response. Promptly returning the call can keep a minor problem from escalating into a major liability.

A home inspection company in the midwest is so conscientious about this that they will actually cancel a new inspection if necessary to make a return call and be the first person back if there is a problem with an inspection.

The Cost of Litigation

It simply costs too much to go to court. The director of a municipal home improvement program in the northeast did a study over ten years of court cases to see if it paid the jurisdiction to go to court. A number of cases were won by the jurisdiction, but there was not a single case where the cost of going to court was less than the cost of settling out of court.

HomeTech's experience supports this. Take for example the cracked heat exchanger in a furnace. The pre-inspection agreement specifically disclaims it, the report disclaims it, but the purchaser knows a contractor who says the inspector should have seen it. A young and hungry lawyer says, "Just give me a $1,500 retainer and I will go to work on it." The cost of a new furnace is perhaps $1,300 and it is quite likely that, if negotiations with the customer do not go well, it is cheaper for the inspection company to buy a new furnace rather than go to court.

Standards of Customer Satisfaction

In today's world, successful companies must meet the customer's perception of satisfaction. Management guru Tom Peters is often asked, "Tom, you mean that if I perform my service according to industry standards, and the customer is not satisfied, that I must take good money out of my pocket to satisfy the customer?" Tom's answer is, "You bet your life!" You must meet the customer's own idiosyncratic, illogical, emotional perception of satisfaction. If you do, you can win the lion's share of business in your industry, because less than 1% of companies today meets this standard.

You may perform inspections to ASHI standards, or the professional standards in your area, and some customers will still think that you should have done better, should have discovered that defect. And you will end up paying for some corrective procedures when you might well have successfully defended a court case. But this level of service is exactly what is going to be necessary for your business to grow and prosper.

The Professional Home Inspector

As a professional home inspector you will present a better business image to the public, and professional conduct will help your defense should a liability claim arise. It is a good practice to hold membership in local organizations such as the Chamber of Commerce, associate membership in the Board of Realtors and the Homebuilders Association, and other civic organizations.

Skills and Knowledge

Home inspectors must have superior skills and knowledge, and their conduct must reflect this superior background.

Usual and Customary: A professional home inspector will follow the usual and customary standards of the industry. This can be a relevant point in a liability case. If the home inspector has followed these practices and company records can prove it, the plaintiff must provide contradictory evidence. A home inspector cannot hide behind industry-wide standards, but they can be an important aspect of defense.

Avoiding liability claims: A professional will base an inspection report on knowledge of materials and methods of construction, including the latest techniques. Reports will not rely solely on disclaimers to prevent liability claims. Professionalism will be evident in performance of the service and in relationships with clients. The home inspector will not advise for or against the client's purchase of a home, but will offer a complete report to aid the client in making the right decision. Under no circumstances should a home inspector offer legal advice of any kind.

Self-protection: A home inspector's skills are the best protection against exposure to liability claims. A professional manner and a complete, properly worded report will offset many so-called "nuisance" claims. If a client should threaten a lawsuit, a strong assertion that the job was done professionally will often discourage the client from pursuing the claim.

Good Business Practices

License: Only one state presently requires a home inspector's license. Never advertise, state, or intimate that you are licensed if you are not. This is dishonest, misleading, and invites liability claims.

Insurance: In addition to usual business insurance, the home inspector should carry errors and omissions insurance. Often this is difficult to obtain; if you cannot get it, your business should certainly be incorporated. If you do have errors and omissions coverage, never advertise it or indicate it to a client. Some people have a hobby of going after businesses that carry large insurance. You do not need to offer invitations to be sued.

238

Incorporation: The advantages of incorporation outweigh the drawbacks. Your attorney will furnish information you need for this decision, but the personal liability aspect is sufficient to convince most business owners in a high liability field to incorporate. Your personal assets are generally protected if your company is incorporated.

If you do incorporate, never refer to yourself as the owner of the business in public relations materials, correspondence, or written reports to your clients. Always sign your name, for example, as John Doe for J. Doe Home Inspection Services, Inc. If you omit the corporate name, you are possibly inviting claims against your personal assets.

Your attorney will handle the incorporation details at start-up, but you should be aware of required procedures. There must be a corporate organization, regular meetings of the board of directors and stockholders, sufficient capital to conduct operations, and correct handling of corporate assets. The president should not treat the assets or employees of the corporation as personal property and should not indicate personal liability or obligation when acting as an officer of the corporation.

Home Inspections and Reports

The home inspection should be conducted in a businesslike manner and the report treated as confidential. The client is the only person entitled to a copy of the report. If the client wishes copies made available to others, such as the seller or real estate agent, you may do so at the client's request.

Be wary of excessive use of disclaimers and warranties in a home inspection report. Some experts advise against them altogether, while others note that a disclaimer or warranty should be written in identical language on all documents in which it appears.

Courts do not look kindly on disclaimers by professionals in the conduct of business. If a company is hired to do a professional job, and the job is not done correctly, no amount of disclaiming can make things right. Warranty programs may cause additional liability problems as well.

Be careful of the wording in a home inspection report. Define terminology such as "good", "fair", or "operational". Never "certify," "warrant," or declare that a system is "acceptable." If the age of a system or appliance is noted, always state it as approximate or give a range of years.

Public Relations and Advertising

Be honest in all public relations materials issued by your company. Avoid puffery and the use of words that misrepresent who you are or what you do.

The right words: Avoid stating or implying that you are licensed unless you really are. It is advisable not to warrant or guarantee anything in a promotional document. A home inspector does not warrant, but only inspects. Have your attorney review promotional materials if you feel the least doubtful about any statement. Do not promise something you may not deliver, like the "straight facts" or a "professional guarantee."

Misrepresentation: Never misrepresent your qualifications, the purpose of a home inspection, or the services you perform. This can be a fast route to a liability claim. Require advance proofs of any printed materials or advertisements. If you use an advertising agency, give your account representative a brief course in home inspections before promotional materials are created.

Contracts

There are three types of contracts: oral, written, and implied. If neither an oral nor written contract exists, the contract is implied.

An oral contract may be made with good intentions on both sides, but each may interpret what was said differently. Also, an oral contract provides no documentary evidence in a court of law.

A written contract spells out what the client can expect from a home inspection: its purpose, what items the home inspector will check, how the report will be presented, and any limitations of the inspection. With a written contract, a home inspector is on a sounder basis should a claim occur.

Implied contracts, without any oral or written elements, are undesirable for both parties. In a legal case, the home inspector would probably be required to establish the burden of proof, not the client.

You may feel that presentation of a pre-inspection agreement will turn away some potential customers. If it honestly presents the purpose and limitations of a home inspection, it should not put off customers. In fact, in HomeTech's experience, no client has refused to sign the pre-inspection agreement.

How to Write a Contract Agreement

The pre-inspection agreement, signed by the client and the inspector, should specify what each party can expect from the home inspection. It should reflect the policies of the company, the laws of the state, and the standards of the industry.

Be precise, consistent, and concise in the wording of your agreement. State exactly what the home inspection will involve and avoid any wording that may be construed as legal trickery to bail you out of a claim. A court will see through this immediately and may refuse to honor any part of a contract that contains such wording. Disclaimers and warranties may come under this heading. Generally a disclaimer, if worded properly and used consistently in all documents, could be helpful in the event of a claim.

Vague or ambiguous wording will be held against you in court. It is your contract and you will be held responsible for its contents.

The agreement should be based specifically on the following:

- Your expertise and experience in the home inspection business
- Standards of the home inspection industry
- A list of what the home inspector is hired to do
- The limits of responsibility in a home inspection

Do not misrepresent any aspect of the inspection or actions to be taken by the home inspector before, during, or after the inspection. Be specific about what you will inspect and how you will inspect it, for example, whether or not actual operation of a system or appliance is included.

Look at the contract from the client's viewpoint and try to imagine what the client could expect from its wording. Do not lead clients to believe you will tell them whether or not to buy the house at the completion of your inspection. Avoid the words "certification," "warrant," and "recommendation" or implications of these words.

If you are not clear on the legal aspects or the format of a contract, consult your attorney. It is extremely important that the attorney review the contract before you put it into final form. A copy of HomeTech's pre-inspection agreement appears in the Appendix.

Contents of a Contract

Title of the Document: You may use CONTRACT or AGREEMENT, or any term that would be legally sufficient.

Date of the Contract.

Names of the Parties: The parties bound by the contract are the client and the home inspection company.

The words, PLEASE READ CAREFULLY: These words should be in bold print at the top of the report. They put clients on notice that ignoring or overlooking any part of the contract is not advisable. The clients should know what they are agreeing to. This notice can present a positive impression to a court in the event of a claim.

Home Inspection Fee: The fee for the inspection and a record of payment, if payment is made on-site, are recommended items to be included with the contract.

Location of the Property: Include the address and a list of the buildings to be inspected. If the property contains detached buildings that will not be inspected, list them and indicate that they will not be a part of the inspection report.

Items to be Inspected: List the components, systems, and appliances included in the inspection. Emphasize that this is a visual inspection and that the home inspection company will not be responsible for latent or concealed deficiencies in any area of the inspection.

Operational Standards: Identify the standards under which you operate. Certain standards are implied because of your professional standing as a home inspector, but be specific about standards such as those of a home inspection association. A copy of these standards should be carried with you and may be attached to the agreement.

Purpose of Report: This can be an important contract clause in a court of law. The purpose of the inspection report would be based on company policy as well as expertise and qualifications in each area of inspection. If you note only **major** defects and deficiencies or your inspection determines only whether a system is functional, indicate this in your agreement.

Limits of Responsibility: Be specific about what your inspection does and does not cover. The agreement should note, for example, that you will not inspect for local code violations; environmental hazards, such as asbestos and radon; and termites. (If you do perform environmental hazard inspections, it would be advisable to use a separate contract and treat it as a separate job.) Also indicate if this is a partial inspection. Leave a blank space for such a contingency, and have both parties initial any additional wording.

Physical Accessibility: Indicate that you inspect only what is physically accessible at the time of inspection. You may wish to note that you do not move furniture, remove carpeting, or scrape paint to determine water stains or wood rot.

Liability Clause: The inclusion of a liability clause may be an effective part of your agreement; discuss this with your attorney and have the attorney write the clause if you decide to use it. If you include the clause in the agreement, also include it in the inspection report. **Use the same wording** in each document.

Entirety of Contract: Indicate that the document constitutes the entire contract. If the client wishes to delete a listed item, both the client and you should initial the deletion. If clients change their mind during the actual inspection and wish you to inspect the item, inform them that, according to the contract, you are unable to inspect it. Inspecting the item after it has been deleted from the contract can create a high degree of liability in a claim. HomeTech does not recommend deleting a listed item.

Disposition of Inspection Report: State that the home inspection report will be submitted only to the client and it is not intended for the use of any other party. This clause can be helpful in the event of third party claims. Indicate how the report will be delivered, such as handed to the client at the close of the inspection or mailed within a certain number of hours.

Signatures: This portion of the agreement will contain blank lines for the client's signature (and spouse's signature, if applicable) and the signature of the home inspector as an officer of the company, if it is incorporated. A proper signature for a corporation would be John Doe for J. Doe Home Inspection Services, Inc.

If the client is unavailable to sign, you may wish to indicate in your telephone conversation that a pre-inspection agreement is standard company policy, but you will waive it in this instance provided the client agrees to it verbally after you give a brief description of its contents. You should also attach a copy of the contract to the written home inspection report. This procedure could make a difference to the court in a liability suit, but don't count on it. A wiser course would be to fax the contract to the client and have it executed before releasing the report.

If only one of two purchasers is present to sign, address the inspection report to that individual only.

Page Numbers: Each page should be numbered in the following manner: Page 1 of 3, Page 2 of 3, Page 3 of 3. As a further protection, you may wish to ask the client to sign each page of the contract. If there are any additions or deletions, or it is amended in any way, each change must be initialed by both the client and the home inspector. A verbal change in any part of the contract would have no standing in a court of law.

Performing the Inspection

The client usually accompanies the home inspector during inspection of the property. Frequently the property owner (the seller) and the real estate agent will be present. Remember who your client is. The other people are not entitled to any information you consider to be part of the confidential inspection report you will submit to your client.

Never volunteer an opinion about any aspect of the inspection. In speaking with your client or answering specific questions, be careful in your choice of words. Follow the wording of your checklist in discussing various components.

If you make a habit of using the correct terminology, you may be better able to defend a misquote.

Follow the checklist of items you are inspecting. Don't let the client or a third party entice you into areas or situations not covered by the agreement and written report. By following the checklist carefully, you may avoid placing yourself into a possibly dangerous liability situation.

Do not be sidetracked by incidental conversation into missing an item on the checklist. Pay particular attention to the roof. Get up on it or view it closely in the manner determined by your company policy. Failure to notice roof conditions can be an expensive mistake.

If you are new at home inspections and don't feel you have sufficient expertise in all areas of the inspection, take an expert with you. You cannot disclaim negligence in any item listed in your contract when you make a mistake because you didn't know any better. As a professional, you are expected to have the necessary knowledge for the inspection, and you may be accused of misrepresentation if you lack it.

Be specific on the checklist about the operation of any system or appliance. If the contract calls for operation and you are unable to perform it for any reason, state this on the checklist and give the reason for nonperformance.

If the client requested a partial inspection and certain items are deleted in the agreement, make a notation to that effect beside each deleted item on the checklist.

In addition to the checklist you use in the inspection, have a notebook handy to jot down information that will not be included in your report. This will include notes for your office files to aid your memory should there be any liability problems.

The following should be included in your personal notebook:

- Names and roles of all those present at the time of inspection
- Weather conditions during the inspection
- Any information volunteered by the seller regarding the house or its components
- Any unusual requests by anyone present

- Any problems of acceptability of any component
- Any noteworthy, troublesome, or unusual situations that occur during the inspection

If you notice a potentially dangerous situation or local building code violation during the inspection, you should bring this to the client's attention and suggest calling an expert in the appropriate field to check it. Or you may suggest that the client request the seller to do so before the final decision to purchase. This information would become a part of your inspection report.

If you are asked to give an opinion about an area outside the field of home inspection, such as environmental hazards or termite presence, suggest that the client call the proper expert.

A picture of the house for your file may be helpful if you are later faced with a claim. The picture may bring pertinent details to mind and help in your defense. Some home inspectors include tape recorders in their equipment list, but if you follow the inspection checklist and keep good notes, the tape recorder could be overdoing your "protective defenses."

Home Inspection Reports

The three basic types of written reports are the checklist, dictated letter, and computer printout. Each of these may be tailored to your company and its operations. The advantages and disadvantages of each are noted in an earlier chapter.

The checklist report may be the safest from a liability standpoint. Each item is listed and the possibility of neglecting an item is reduced. The checklist format also would follow the listing in your pre-inspection agreement and thus provide more specific documentation in the event of a liability claim.

If you prefer a more personal touch but do not want to write a personal letter for each report, a combination of the checklist and written report can be achieved by leaving several blank lines after each main category in the checklist. This gives you the opportunity to fill in the blanks with specific information about any items in that category. The blank lines could be headed "Additional Information" or "Comments," or a title of your choice. If you have nothing to write in the blanks, an indication of "None" would be acceptable.

The report should be specific in every detail and conform closely to the wording of the pre-inspection agreement. If a disclaimer is used, the wording in the report should be identical to the disclaimer in the agreement. Using your knowledge and expertise, you base the report on the facts as you determine them. An opinion is not part of the package deal, nor is a recommendation. You suggest, not recommend; you report, not advise; an item is "functional" or "operational", not "acceptable".

Do not use a specific figure in estimating cost of repairs; give an approximate figure or range of figures if you feel that estimating a cost is pertinent. The same rule applies to the life expectancy of a system or appliance.

Suggest that the appropriate specialist be called if you feel an item requires expert attention. This is particularly important if you believe there may be serious structural problems or dangerous situations. It is better to be overcautious than to minimize a potential problem which could result in negligence on your part.

You may be asked why you are recommending an expert and why you can't analyze the problem yourself. Remind the client or agent that you are a generalist. Your job is to recognize the existence of a problem and then, if necessary, recommend a specialist to determine the solution.

Your responsibility to the client is to gather all the facts available for each component included in the contract agreement and furnish these facts to the client. The written report reflects this responsibility and its completion.

Your report may be a basic part of the client's decision whether or not to purchase the house, but it should be a statement of observable facts and never include your personal opinion as to what the client should decide. That opinion could leave you vulnerable to a liability claim of major proportions and should be scrupulously avoided in every verbal and written communication with your client.

Be especially careful in describing your results verbally to the client. This is another advantage of the checklist report form. If time is of the essence and the client is required to make a quick decision, the few minutes it takes to complete the checklist report at the site would assist in a timely decision and protect you from possible liability if the client misinterpreted or ignored any facts in an oral report.

Writing the Report

Use a checklist, whether your report is in checklist or narrative form. The checklist should follow the list of items contained in the pre-inspection agreement.

Be specific about each item that was inspected. Indicate if a system or appliance was operated. If not, give the reason for non-operation. If any items were deleted from the contract agreement prior to inspection, so indicate in the report.

If it is your professional opinion that any component should be checked by a specialist in that particular field, specifically indicate this in the report.

If you estimate the age of the house, a system, or an appliance, use the word "approximate" and give the age in a range of years, rather than a specific year.

If you use any terms such as "good," "fair," "poor," "operational," "functional," define the terms used. Be careful of stating that a system is in good condition; the client may believe you mean it will last for years, and you could invite trouble if you inadvertently imply that it will, even though that was not your meaning.

If an item in your contract agreement or in the checklist report does not apply to the inspection, indicate it in the report and give the reason why. If you use a narrative report form, include this information.

Do not generalize. Indicate only the facts as obtained from your inspection.

Use wording consistent with that in the pre-inspection agreement.

Put a disclaimer in bold type at the beginning of the report if you include it anywhere. Do not try to hide it in small print on the last page.

Follow industry standards in your reporting procedure. If you are a member of a home inspectors association, indicate that you are and that you use that association's standards.

If you were asked to inspect an item not usually covered by your company and you do not feel qualified to do so, indicate this in your report and suggest the proper type of specialist for the inspection.

State in clear language that your inspection does not include environmental hazards, termite presence, or hidden structural defects. This is doubly important if you are a professional engineer or registered architect. If there is a liability claim, a court may hold these professionals to higher standards than most home inspectors would be held.

If you are a professional engineer or registered architect, sign a normal home inspection report with your name only, "John Doe." Putting your credentials on the report only raises your liability.

If you have reason to believe there are structural defects in the house, indicate this in your report, along with the suggestion that a structural engineer be consulted.

If you are qualified to perform an environmental hazard inspection, make the inspection separate from the regular home inspection. Use a separate contract and reporting form and handle the job as a separate service to the client, and use a separate pre-inspection agreement.

Indicate that your home inspection report completely covers all aspects of the inspection described in the pre-inspection agreement. Do not leave any holes in your report. If an item is not applicable, say so and explain why.

As in the contract agreement, always sign the report as "John Doe for J. Doe Home Inspection Services, Inc.", if your company is incorporated. Do not take unnecessary personal responsibility for the actions of your company.

Disposition of Report

The report is given or mailed to your client only. A copy is kept in your files, along with any additional notes you made for your personal reference. If a client requests that a copy of the report be sent to a third party, obtain a written, signed request from the client before making the copy, or perhaps suggest the client make the copy and forward it. Your sole obligation is to your client. It is even acceptable business practice to stamp the word "Confidential" on each page of the report.

Although you probably will receive most of your referrals from real estate agents, keep in mind that in almost all cases the agent represents the seller and not the buyer. Since most of your clients will be buyers or potential buyers, you would not logically or legally have any reason to report to the broker. Be aware of any possible appearance of collusion between your home inspection business and real estate companies.

If an agent wants a copy of the report, recommended procedure is to ask the purchaser at the time of inspection if it is all right to send the agent a copy of the report. Some inspection companies require this permission to be in writing.

Unless the inspection report has been hand-delivered to the client at the inspection, you will send the report to the client within a specified time with a cover letter. Do not attempt to summarize the report in the letter, and do not make general statements or opinions about the condition of the house. Merely transmit the report with a brief mention of the date of inspection and perhaps the time involved if the client was not present, and thank the client for the business.

Disclaimers and Warranties

Disclaimers and warranties have been viewed by some courts as unprofessional practice. They may harm rather than help the defense of a liability suit, especially if improperly worded or not included in the contract agreement, if there is a written contract. A disclaimer should be included in the contract or the client should receive ample notice of the disclaimer prior to inspection.

Some home inspection companies offer a warranty. If you decide to do this, review the intent and wording of the warranty with your attorney. It is not good practice to include the warranty in advertisements or other public relations materials. Your attorney can advise you of possible legal implications.

Liability Claims

Professionalism, knowledge, and expertise in the home inspection field are the personal assets that will help you to prevent liability claims against your company. Moreover, these same traits will constitute your strength and defense against claims should they arise. A list of the most commonly experienced claims appears in the Appendix.

As owner of the company, you are likely to be most careful during an inspection and unlikely to take chances. It is important that any additional inspectors you recruit be tied into the liability on each inspection, to give them maximum incentive for care and thoroughness. When it's a hot summer day, and the attic crawl space is 140 degrees, an inspector not tied into liability might not check that space for flashing problems or roof leaks.

Steps to Take

Your response to a call about a problem should be immediate. First, offer to set a time to come back to the house for a re-check, look at the condition and see what it involves. Let the caller know right away that you are willing to come out at no charge and consult with them about the problem. Your goal is to allay their fears and prevent their calling someone else for a second opinion.

If your company has more than one inspector, the first choice for a re-check would be the person who did the original inspection. If the customers do not want that inspector back, send the best-qualified inspector you have, preferably the company owner.

Most of the time, this quick response will solve the problem. Usually the problem turns out to be normal maintenance that the purchasers did not understand, or a condition that was reported as minor in the inspection report. It is surprising how seldom purchasers really read and absorb the information in the report. So the result is, the purchasers are delighted that you have come out, their questions are answered, they understand more about their house, and your professionalism and customer satisfaction are increased.

Sometimes a small problem can be resolved with no expense or by paying for a service call as a goodwill gesture. A typical example is an electric outlet that does not work, and all the customer wants is reimbursement of the electrician's $50 charge. It may be well worth it to pay this amount, achieve customer satisfaction, and forestall further problems.

Occasionally there will be some question about liability that requires further study. In such cases, the inspector at the re-check tells the purchasers that the company will send a letter within a week saying what will be done. Usually this letter will show how the pre-inspection agreement clearly covers the subject, and that the company has no liability for any corrective measure. But the final paragraph of the letter should read something like this:

"Our company takes pride in our reputation, and we want to make every effort to be fair. Therefore, if for any reason a client is not happy with our inspection, we will refund the inspection fee."

HomeTech's policy is to refund the original inspection fee when there is dissatisfaction, almost no matter what the cause.

Here is an example of language used in a release of claims:

Sample Release of All Claims

For and in consideration of the payment to me of the sum of $_____ and other good and valuable considerations, the receipt and sufficiency of which is hereby acknowledged, I, [name] of [address] have released and discharged, and do by these presents, for myself, my heirs, executors, administrators and assigns, release, acquit and forever discharge [home inspection company], [inspector's name], and all other persons, firms, and corporations, whether herein named or referred to or not, of and from any and all past, present, and future actions, causes of actions, claims, demands, damages, costs, loss of services, expenses, compensation, third party actions, suits at law or in equity, including claims or suits for contribution and/or indemnity, of whatever nature, and all consequential damage on account of, or in any way growing out of any and all known and unknown personal injuries, death, and/or property damage resulting, or to result, from and inspection that occurred on or about [date of inspection] of the property located at [address of inspection].

I understand that the payment in connection herewith is not to be construed as an admission of liability on the part of the persons, firms, and corporations hereby released by whom liability is expressly denied.

This Release contains the entire agreement between the parties hereto, and the terms of this Release are contractual.

I further state that I have carefully read the foregoing Release and know the contents thereof, and I sign the same as my own free act.

[Name]	Date

If a customer insists that there is more liability than just the inspection fee, the next step is to take the depreciated value of the condition. For example, a roof was 14 years old and showing signs of deterioration. The inspector said the normal life of the roof was 15 to 20 years, and a new roof should be budgeted for in the next one to five years. It happened that the roof was leaking at the time of inspection, this was clearly visible, and the leak was not reported. If the normal life was 20 years, about 1/3 the life was left, so the company offers to pay 1/3 the cost of a new roof. This is usually accepted readily by customers.

In some cases, the property condition clause of a sales contract leads customers to refuse the depreciated value offer. A good example is a cracked heat exchanger. The customer's point is that if the heat exchanger problem had been discovered at the inspection, the seller would have been required to provide a complete new furnace of equal quality, so the customer wants the cost of a new furnace.

Often a client will call with a complaint and threaten a lawsuit before taking any definite legal action. This is where you put your professionalism to work. Obtain all the information you can from the client and, if possible, visit the site and obtain photographs.

Your explanation about how you gathered the facts concerning the item in question, as described in the contract agreement and according to standards of the home inspection industry, frequently will help the client to understand you are not responsible for the problem. A friendly and helpful approach and a sincere belief that you performed the inspection properly may be all that are required to disperse the client's threats.

Be careful in your statements to the client, however. Make no admission of negligence or any other statement that subsequently may be used against you. If you have errors and omissions insurance, do not mention it to the client. Often the word "insurance" triggers dollar signs in front of a person's eyes.

If you feel the claim may have merit -- home inspectors, like everyone else, can make mistakes -- an offer of a monetary settlement or replacement of the item could prevent future legal action. Before making an offer, you may wish to consult your attorney about the nature of the settlement and the proper release form to prevent further liability claims.

Defense of a Liability Claim

Should your efforts to resolve the problem fail and the client persists in the claim, immediately notify your insurance carrier if you have errors and omissions insurance. Be familiar with the carrier's instructions on notification of claims and how to avoid creating additional problems through your own actions. Give the claims agent all the information you have obtained from the client about the claim.

If you carry E&O insurance, your defense will be handled by the insurance carrier's attorneys. It is a good idea, however, to notify your own attorney of the claim, and if you have no insurance it is imperative that you consult your attorney immediately.

Be honest with your attorney and furnish all the facts in the case: names and addresses of those involved, your own actions during the inspection, your attempts to resolve the case, information obtained from the client, how you feel about the possibility of liability on your part, and other pertinent details.

If you feel settlement is the best course and you have tried to settle with the client, perhaps your attorney will be successful in dealing with the client's attorney or directly with the client if there is no attorney.

Your attorney can advise you about how to proceed with what you believe is strictly a nuisance claim. Whether you wish to make a small settlement just to rid your company of the claim may involve your business judgment and principles and should be discussed with your attorney.

The records in your file can be your best defense in any liability claim. Now is the time to remember all details of the case. The notes you made about the inspection will be helpful to your memory, as well as providing documentary evidence of your actions and those of others. A word of warning: anything in your files not directly addressed to your attorney might be subject to discovery and have to be turned over to the plaintiff should the case proceed toward trial.

If this is your first liability claim, you may learn several lessons that could make the trauma, expense, and inconvenience worthwhile. These may include the results of hurried, sloppy, or incomplete inspections; the danger of oral reports or casual conversations with the client that contain sentences beginning with, "It is my opinion . . .," or "I believe you should . . ."; conversations with third parties, such as the real estate agent, seller, or other persons present during the inspection, or furnishing them copies of the inspection report.

If you are guilty of none of these, you will learn the importance of well-written contracts and reports, a professional attitude during your inspections, and an efficient and complete filing system.

Third Parties

Third parties may file claims against your company or they may share some of the responsibility for liability in other claims. Generally the third parties a home inspector will encounter are real estate brokers and salespeople and the sellers of the property being inspected.

Third Party Claims: The real estate company or the seller could sue you for loss of the sale of a property if they believe your home inspection was in error, or if they believe your recommendations or opinions drove away the buyer. The business relationship between the home inspector and the client is confidential and should involve no other parties; however, a broker or seller could attempt to obtain confidential information to pursue a claim. This possibility points up the care you should use in your conversations with the client in the presence of others, and the confidential treatment of your home inspection report.

If you receive a telephone call about the inspection report, be sure that you are speaking with your client and not someone else. Send the report and cover letter to your client only and be sure you have the correct name and address.

Some lawyers believe that third party claims against home inspectors will be more frequent in the coming years. However, as of this writing HomeTech has never been the subject of a third party claim by a seller or real estate agent.

Third Party Defendants: If you are sued by a client, it may be possible to bring in other parties to assist in defense of the claim and to share responsibility for any payments ordered by the court. You may have based portions of your report on information offered by the seller, such as dates of purchase of certain components. Although you should not have relied on such information, you may be able to bring in the seller as a codefendant through a cross-claim.

Other possible codefendants may include construction or repair people who were called subsequent to your inspection, as suggested in your report. If you can prove the claim arose because of their negligence, rather than yours, you would be in a position to transfer all of the liability to the third parties.

Avoiding Liability Claims

Do not make mistakes or try to cover up your lack of knowledge or expertise in any area. If you feel unqualified to handle a job, refer the client to someone else or bring in an expert.

Inform the client of what a home inspection will include. The client may have totally different expectations of what an inspection entails. A well-written contract agreement can be an excellent educational tool, but be sure the client understands what the contract says. Never pressure a client to sign the contract agreement. If the client seems uncertain or unwilling to commit to a contract, you would be wise to go on to the next client.

Make certain that every item in your agreement is covered in your report. If you use a checklist report form, do not leave any blank spaces. Clients may easily assume a blank space means the system is in good condition.

Base your inspection and report on industry standards. Be aware of local government disclosure or competency regulations. Be certain your contract agreement, inspection, and report conform to these regulations.

Do not attempt to include any extras that do not belong in your inspection, such as a check for environmental hazards or possible hidden structural defects. Be explicit in your contract agreement that a home inspection does not include these items.

Be careful of your choice of words. Review any disclaimers or warranties with your attorney. Define words that you use in describing the condition of the roof, systems, and appliances. Emphasize that this is a visual inspection only, and list the actions you do not take, such as pulling up carpet, moving furniture, or scraping paint.

Be honest and professional in your advertising and public relations materials. Do not make any promises or guarantees you cannot honor. Promises and guarantees are not part of a home inspection and do not belong in public relations materials, contracts, reports, or any other aspect of the business.

Negligence

The burden of proof is on the plaintiff, but negligence is a charge with serious ramifications, particularly if the plaintiff is able to prove fraud, malice, or gross negligence. Any of these could result in punitive damages in addition to liability. Most E&O insurance does not cover punitive damages. Also, the court can direct payment of these damages by the individual rather than the corporation.

Presence of Negligence

Negligence is defined by Blackwell's Dictionary of Law as: "The omission to do something which a reasonable man, guided by those ordinary considerations which ordinarily regulate human affairs, would do or the doing of something which a reasonable man would not do."

To determine the presence of negligence, a court would substitute the word "professional" for "reasonable man" in the definition above. There are four legal requirements for negligence:

- ✓ A duty or obligation recognized by the law requiring a certain standard of conduct for protection of others against unreasonable risks.

- ✓ Failure on a person's part to conform to the standard required, or a breach of duty.

- ✓ A reasonably close causal connection between the conduct and the resulting injury.

- ✓ Actual loss or damage resulting to the interest of another.

A "reasonable" home inspector presumably has superior knowledge, skills, and intelligence and acts consistently with this superior background. This would include:

✓ Accurate facts as opposed to inaccurate deductions.

✓ Good professional practices -- what is customary and usual in the profession.

✓ Use of industry-wide standards.

Following accepted standards would not necessarily prove that you were not negligent, nor does it prove that you were negligent if you deviated from them. But your defense may be stronger if you can prove that you followed the standards. The plaintiff would be required to prove that both the standard and the home inspector were in error.

Causal Relationship

The plaintiff may be required to show a causal relationship between the defendant's negligence and the damages that resulted. If you believe there is no causal relationship between your actions as a home inspector and the resultant damages of the buyer, consult your attorney regarding the possibility of this defense.

Appellate Court Cases

Home inspection is a new industry and we have only three liability decisions involving home inspections from the appellate courts. Home inspectors should be aware of these rulings and perhaps use that knowledge in revising their business practices. Also, they should know that lower courts must follow rulings of their respective state appellate courts. These rulings, therefore, may be very important to home inspectors operating in New Jersey, Connecticut, and Indiana.

New Jersey Superior Court

This case was appealed by the defendant home inspector before a lower court trial on grounds that it should be thrown out of court. A home buyer's roof leaked after 8-12 months of purchase, although the home inspector's report had given the roof a 15-40 year life expectancy at the time of the inspection. The buyer sued the home inspector for negligence. The Superior Court found against the defendant and sent the case back to the lower court for trial.

The home inspector's brochure stated the buyer will be told "the straight facts," home inspectors are "expert advisers," and home inspectors are informed "as to the latest developments in home construction and maintenance". This brochure led the buyer to believe the inspection report concerning life expectancy of the roof. (Note: The brochure has found its way into numerous lower courts as Exhibit A.)

The court rejected the clause in the pre-inspection contract which limited the home inspector's liability to the cost of inspection and stated the contract should be more specific about what the plaintiff could have expected if the home inspector was negligent. Also, the court stated the client should have had the right of negotiation regarding this clause.

The court rejected a warranty offered in the inspection report by the home inspector, upon payment of a pre-inspection fee, to fix anything that was wrong with the house. The court declined to comment on the reasons for this finding. Courts generally hold that a professional would be negligent if unqualified for the job.

Connecticut Superior Court

Several serious problems, including previous fire damage, were missed by the home inspector, who claimed the problems were in existence before his inspection, so he was not liable. He claimed there was no causal relationship between his alleged negligence and the damage. The court found for the plaintiff.

Breach of contract (implied) was involved. The buyer relied on the home inspector and his report. The contractual relationship between the home inspector and buyer was breached by the home inspector's neglect to locate the problems. There was no written or oral pre-inspection agreement or contract in this case.

The court found the home inspector misrepresented his skills and ability to perform the inspection by implication when he agreed to handle the inspection.

Indiana Court of Appeals

A buyer saw the home inspector's advertisement and also received a recommendation for the home inspector from a real estate company who hired the home inspector. Repairs noted on the inspection report were performed before the buyer moved in. Later, the buyer found other severe problems, including a bad roof, electrical and plumbing problems, not listed in the report. The court found for the buyer, a third party beneficiary.

The plaintiff relied on the home inspector's advertisement, in which he mentioned his own name as owner and claimed he was a licensed home inspector. The fact that there was no licensing law in the home inspector's state, among other findings, resulted in the court piercing the corporate veil, and the home inspector was required to pay punitive damages, in addition to other damages, from his personal funds. The court also found that the corporation had neglected to issue stock, which was in violation of state law.

Summary

It is important for you to have a clear understanding of the ramifications of your professional actions. You must be careful in what you inspect, what you report, and the form you use for reporting it. Any slip-ups could cost you your job, cost your employer thousands of dollars, or if you're self-employed, you could be ruined financially by the liabilities you create as a home inspector.

You have assumed the role of a professional, an expert, and you must live up to society's expectations of professional people. Your product is information and education. The quality of that information will make the reputation of your company.

Among the business practices you will need to consider are whether to incorporate and whether to carry errors and omissions insurance. In the past,

E&O has been expensive and hard to come by, but the picture can change rapidly. In general, self-employed home inspectors can gain a degree of personal liability protection by forming a legitimate corporation and then acting carefully as an employee of the corporation in all professional dealings.

Always keep the corporate shield between your personal assets and the world at large. Always sign your name as for your corporation. Never present yourself as owner and never make public the fact that you carry E&O insurance.

In your advertising and brochures, don't make exaggerated claims. Don't call yourself "licensed" if you are not, and don't offer "expert opinions" or "professional guarantees." You can offer fast service, accurate reporting, and factual reporting of conditions visible to the trained observer on the day of the inspection. You can offer to give the clients a useful planning tool for the purchase and ownership costs of the house. You can and should perform inspections consistent with the standards of the home inspection industry, whether or not you are a current member of any national or state organization.

You should use a written contract agreement. In performing the services you have agreed on, don't ever forget the person for whom you are working -- and don't offer information to anyone else unless requested to do so by your client.

The three types of reports you might make are checklist, narrative, and computer print-out. While each has advantages and disadvantages, it is a safe practice to use a checklist for making your actual inspection. That way you train yourself not to overlook any items in the conduct of your daily business. Follow industry standards and do the best job you are capable of doing. In this way you will believe in your work product and if the time comes to defend your past actions, you'll be confident and prepared.

√ Keep good notes.
√ Believe in yourself.
√ Believe in, and be forthright with, your attorney.
√ Report facts, not opinions.

Always give a range of years for age or condition of components, and a dollar range for cost of upkeep and repair. Never be specific.

If you are pressed to give an answer you can't be comfortable with, call for an expert and make it clear in your report that such is your recommendation. Then leave the question to the expert; don't try to answer it yourself.

Review your strategy with your attorney early on if it appears that a claim is going to be placed against you, even before you offer to settle. Be careful with your advertising claims, your inspection reportage, your business structures, and your selection of an attorney.

Beyond that, **GOOD LUCK!**

Inspection Schedule Card

Date Inspection Scheduled	Day of Week	Time	Building Analyst

Purchaser _____

Present Address _____

City, State, Zip _____

Home Phone	Office Phone	Office Phone

	Day	Date	Time	Real Estate Agent	Company
Contingency expires	_____	_____	_____		
Report required by	_____	_____	_____		
Type of report ☐ B.A.R. ☐ 404 ☐ Narrative ☐ Partial				Home Phone	Office Phone
☐ Other:					
Payment by ☐ Check ☐ Credit Card ☐ Send Bill					

_____ | Area |

PPHA Address

Invoice No.

Special instructions, directions to inspection site, etc.

Price of Building

Inspection Fee

Source of Call

Who made Call

Owner	Who will be present:		Call Taken by
	☐ Purchaser	☐ New House ☐ Vacant	
	☐ Agent	☐ Used House ☐ Utilities not on	
	☐ Owner	☐ Multi-Family	Date
Phone No. at Building	☐ _____	☐ Condo Unit ☐ _____	
		☐ Commercial	

FORM 426 ⓒ 1981 Home-Tech, Inc.

263

Pre-Inspection Agreement

PRE-INSPECTION AGREEMENT
(PLEASE READ CAREFULLY)

Between: HOMETECH SYSTEMS, INC.

_____
(Inspector)

And: _____
(Customer)

Re: _____ $ _____
(Property Address)   (Fee)

COMPANY agrees to conduct an inspection for the purpose of informing the customer of major deficiencies in the condition of the property. The inspection and report are performed and prepared for the sole, confidential and exclusive use and possession on the CUSTOMER. The written report will include the following only:

- structural condition and basement
- electrical, plumbing, hot water heater, heating and air conditioning
- quality, condition and life expectancy of major systems and appliances
- kitchen and appliances

- general interior, including ceilings, walls, floors, windows, insulation and ventilation
- general exterior, including roof, gutter, chimney, drainage, grading
- estimates on repairs and probable expenses for the next 5 years

It is understood and agreed that this inspection will be of readily accessible areas of the building and is limited to visual observations of apparent conditions existing at the time of the inspection only. Latent and concealed defects and deficiencies are excluded from the inspection; equipment, items and systems will not be dismantled.

Maintenance and other items may be discussed, but they are not a part of our inspection. The report is not a compliance inspection or certification for past or present governmental codes or regulations of any kind.

The inspection and report do not address and are not intended to address the possible presence of or danger from any potentially harmful substances and environmental hazards including but not limited to radon gas, lead paint, asbestos, urea formaldehyde, toxic or flammable chemicals and water and airborne hazards. Also excluded are inspections of and report on swimming pools, wells, septic systems, security systems, central vacuum systems, water softeners, sprinkler systems, fire and safety equipment and the presence or absence of rodents, termites and other insects.

The parties agree that the COMPANY, and its employees and agents, assume no liability or responsibility for the cost of repairing or replacing any unreported defects or deficiencies, either current or arising in the future, or for any property damage, consequential damage or bodily injury of any nature. THE INSPECTION AND REPORT ARE NOT INTENDED OR TO BE USED AS A GUARANTEE OR WARRANTY, EXPRESSED OR IMPLIED, REGARDING THE ADEQUACY, PERFORMANCE OR CONDITION OF ANY INSPECTED STRUCTURE, ITEM OR SYSTEM. COMPANY IS NOT AN INSURER OF ANY INSPECTED CONDITIONS.

It is understood and agreed that should COMPANY and/or its agents or employees be found liable for any loss or damages resulting from a failure to perform any of its obligations, including but not limited to negligence, breach of contract, or otherwise, then the liability of COMPANY and/or its agents or employees, shall be limited to a sum equal to the amount of the fee paid by the CUSTOMER for the Inspection and Report.

Acceptance and understanding of this agreement are hereby acknowledged:

_____   _____   _____   _____
Company Representative   Date   Customer   Date

Form 414 © 1988 HomeTech, Inc.   HomeTech, Inc.

Sample Record Keeping Forms

Pro Forma Operating Budgets

Operating Revenue:	$50,000	$100,000	$200,000	$500,000
Direct Costs @ 40%	20,000	40,000	80,000	200,000
Operating Expenses:				
General management 5-10%	2,500	5,000	20,000	50,000
Secretarial, incl. fringes 6%	3,000	6,000	12,000	30,000
Rent, incl. utilities 2%	1,000	2,000	4,000	10,000
Office equip., supplies 4%	2,000	4,000	8,000	20,000
Telephone, incl. 1/2 cellular 2%	1,000	2,000	4,000	10,000
Telephone answering service	1,000	2,000	--	--
Accounting/legal retainers	600	1,200	2,400	3,600
General insurance 1-2%	1,000	2,000	4,000	10,000
Education, seminars, conventions, books	500	1,000	2,000	5,000
Advertising, all types 10%	5,000	10,000	20,000	50,000
Car expense	5,000	5,000	6,000	8,000
Credit card expense 1%	500	1,000	2,000	5,000
Claims and legal expenses 5%	2,500	5,000	10,000	25,000
Total Operating Expenses	25,600	46,200	94,400	226,600
Net Profit	$4,400	$13,800	$25,600	$73,400

Appendix

Subcontractor Record

NAME _____

ADDRESS _____

_____

OFFICE PHONE _____ HOME PHONE _____ PAGER _____

YEARS IN BUSINESS _____ NO. OF INSPECTORS _____

LICENSE NO._____

INSURANCE:

Type: _____ Carrier: _____

_____

ASSIGNMENTS

Date Assigned	Site/ Customer	Date Completed	Fee	Date Paid	Comments

Start-Up Funding

INITIAL EXPENSES (one time only)
Equipment
     Field $_____
     Office (including installation costs)   _____
Office remodeling and decorating   _____
Public utility deposits and installation expense _____
Legal fees   _____
Other professional fees   _____
Licenses and permits   _____
Advertising and promotion for opening   _____
Cash for unexpected expenses   _____

TOTAL INITIAL EXPENSES $_____

ESTIMATED MONTHLY EXPENSES (Based on estimated yearly sales volume of $_____)

	Monthly Expenses	No. of Months	Amount Required
Salaries and Wages			
Owner/manager			
Office staff			
Field staff			
Subcontractor fees			
Rent			
Advertising and Marketing			
Insurance			
General Liability			
Workmen's Compensation			
Motor vehicle			
Errors and Omissions			
Other			
Telephone			
Other utilities			
Supplies			
Taxes			
Travel expense			
Credit card expense			
Interest			
Maintenance			
Legal fees			
Accounting fees			
Miscellaneous			

TOTAL RESERVE FOR MONTHLY EXPENSES $_____

TOTAL START-UP CAPITAL REQUIRED $_____

Employee's Earnings Record

NAME: SOC. SEC. #:

POSITION: DEPARTMENT:

EMPLOYEE NO.:   PAY RATE:

Earnings: Deductions:

Pay Per. End.	Reg.	O.T.	Total	Cum. Total	FICA	Fed. inc. tax	St. inc. tax	Life ins.	Total Ded.	Net Pay

Business Goals

For year ending December 31, _____

Goal (By Priority)	Strategies	Est. Date of Achievement	Results
> Three-Month Goals:			
To start business			
To reach $2,000 monthly sales			
To contact 25 real estate brokers			
> Six-Month Goals:			
To reach $6,500 monthly sales			
To open office			
To hire secretary			
To establish direct mail promotion			
> One-Year Goals:			
To reach $12,000 sales volume			
To subcontract 25 percent of inspections			
To make four association presentations			

Appendix

Worksheet for Tax Obligations

Tax	Due Date	Amount Due	Pay to:
FEDERAL:			
Employee income tax, Social Security tax			
Owner-manager's and/or corporation income tax			
Unemployment tax			
STATE:			
Employee income tax			
Owner-manager's and/or corporation income tax			
Unemployment tax			
Other			
LOCAL:			
Real estate tax			
Personal property tax			
Licenses			
Other (if applicable)			

Writing A News Release

When preparing your press release, whether it is to report a development or event, or a feature release for general use anytime, follow an established format. The editor receives stacks of publicity releases and has more information than time, so make your submission easy to use. This will improve your chances of seeing the item in print. Here are some simple guidelines:

√ Always double-space or triple-space your text and leave wide margins on both sides.

√ Skip an extra line between paragraphs. Leave at least one inch of white space all around the text. This gives the editor space to mark the copy for publication or make minor editorial changes.

√ Use plain white paper, not your business letterhead. Place your name and telephone number in the upper right corner next to the word, "CONTACT:" Always use a release date and label it clearly. This should appear directly below the contact identification. It signals to the copy editor whether there is an element of timeliness involved in the material and saves a search through the text.

√ For announcement of an event to come, set the release date several days before the event. If you are supplying feature material, label the release date as follows: "FOR RELEASE: At will."

√ Whenever a release runs more than one page, leave a space at the bottom and center the word "more" at the bottom. At the end of the release, place the word "end" in the same position.

The news style. The average newspaper article is designed to present the most important facts as soon as possible, with each successive paragraph declining in importance. This is the inverted pyramid style, used for the simple reason that news space is continually being cut. If the most vital information is in the first two or three paragraphs, chances are the story will not be crippled by cutting from the bottom up.

Use the five W's formula in the lead paragraph. That is, tell immediately who, what, when, where, and why. You should be able to do that in no more than three paragraphs. If you have more space, then you can elaborate the "why" with some "how". For example:

> "John Phillips will show slides of structural dangers in local residences on Thursday night at the Town Library, to warn home-owners how to prepare for the coming storm season.

> "Phillips, a professional home inspector in this area for the last three years, has compiled a list of twelve symptoms of danger. Your home may be trying to signal serious illness, and you could be missing the point.

> "Pictures of actual conditions show the consequence of ignoring the signs in extremely graphic detail. Phillips' slide collection includes before and after shots of houses that were repaired and withstood the hurricanes of last year, along with those that were ignored until the storm caused their collapse.

> "His library contains well over a thousand slides which he took in the course of his regular professional inspection rounds. Phillips has presented these documentaries at civic groups and before the Home Preservation League almost monthly for the past year . . . etc."

Upon examining the four paragraphs above you can discover that while it might be nice to have the whole text printed, the most important facts are preserved if just a two-paragraph blurb gets into the paper. You may find it helps to write out a list of the five W's before writing the release.

As a home inspector familiar with writing reports, you won't have any problem writing the sort of material needed for a useful home maintenance and inspection series. Just think through your subject matter and outline the article before you begin to write. In a feature article, the inverted pyramid structure is not normally used. You can write in report style. Just remember not to exceed the word length you have been assigned.

Job Description for Secretary/Office Manager

The secretary/office manager performs and administrative duties for the company, handles operation of the office on a day-to-day basis, and is the personal contact between members of the public and company employees and subcontractors. This job description assumes the company does not employ a bookkeeper.

OFFICE HOURS

The office is open Monday through Friday from 8 a.m. until 5 p.m. The secretary/office manager is expected to work 40 hours per week. Normal hours would coincide with office hours, with one hour for lunch. These hours may be somewhat flexible depending on company and individual requirements.

RESPONSIBILITIES

The secretary/office manager is responsible for the following office functions:

Answering the office telephone.

Scheduling appointments with customers and obtaining required information for home inspections.

Maintaining contact with the home inspectors, verifying they are informed of their appointments, and controlling the office communications system including beepers, answering service, and cellular telephones.

Maintaining the company checkbook, financial and accounting records, and sales, customer, and subcontractor records.

Typing inspection reports, correspondence, and other documents, as required.

Handling incoming and outgoing mail and verifying that inspection reports are submitted to customers in a timely fashion.

Assisting in the design and development of marketing aids.

Maintaining the office equipment and handling service requirements.

Preparing checks in payment of invoices, subcontractor fees, and other expenses as directed.

Maintaining all record-keeping systems and files in an up-to-date manner.

SPECIFIC DUTIES

Incoming telephone calls:
- Screen potential customers and request necessary information.
- Answer customers' questions about home inspections.
- Schedule appointments.
- Verify that appointments have been kept by home inspectors.
- Maintain telephone log and lead records.

Customers:
- Maintain customer files.
- Handle fee payments, make up deposit slips, and deposit checks in bank; request credit card approval, if required.
- Type inspection reports, cover letters, and other correspondence.
- Prepare invoices for customers not paying at time of inspection and contact customers not paying invoices in a timely fashion.

Home inspectors/subcontractors:
- Maintain schedule for each.
- Prepare and maintain cost and payment records for each inspection.
- Control communications network between inspectors and office.

Record keeping:
- Maintain bookkeeping system
    - Payroll
    - Cash journal
    - General ledger
    - Accounts payable, receivable, and other ledgers, as required.

- Maintain sales records.
- Maintain customer and subcontractor files.
- Maintain other records as needed.

Marketing aids:
- Assist in development of presentations to real estate and civic groups.
- Prepare, or assist in preparation, of brochures, direct mail, and other advertising material, as directed.

Office equipment and supplies:
- Purchase necessary equipment and supplies.
- Handle service calls.
- Maintain inventory of equipment and supplies.

Reports: (Prepare or assist in preparation, as directed)
- Financial (monthly, quarterly, annually)
- Sales and leads
- Tax
- Others, as directed

Job Description for Bookkeeper

The bookkeeper maintains the following records as indicated. All records are to be kept up to date and reports submitted in a timely and accurate manner. Double entry ledgers and records are to be established and maintained according to acceptable bookkeeping standards and procedures.

If, for any reason, the bookkeeper is unable to maintain the records or furnish reports to management in a timely manner, it will be the responsibility of the bookkeeper to alert the president/owner of the situation, give the reasons for it, and request assistance, if required.

All company financial records are confidential, and this confidentiality shall be respected by the bookkeeper.

Cash Receipts:
- Maintain record of all cash receipts.
- Prepare deposit slips and deposit receipts daily.
- Post receipt of customer fees to appropriate customer files.

Payroll:
- Produce weekly payroll for salaried employees and subcontractors.
- Maintain payroll and attendance records in a separate file for each.
- Maintain fee and payment records in a separate file for each sub-contractor.

Accounts Payable:
- Record, by vendor and invoice number, all payments due.
- Request appropriate approval for payments due.
- Pay all invoices in a timely fashion and take advantage of all discount dates.
- Record recurring expenses and pay promptly.
- Post payments to appropriate ledgers and accounts.
- Use serially numbered checks for all disbursements except petty cash.
- Remove signature section of voided checks and maintain in file.
- Submit all company checks for signature to owner or other authorized person.

Accounts Receivable:
- Maintain accounts up to date.
- Submit periodic report of outstanding receivables to management.
- Deposit all receipts daily.
- Make copy of incoming check and deposit slip for customer file.

Cash Flow:
- Prepare a daily report of bank balances, receipts, and disbursements for management.
- If a cash shortfall is anticipated, notify management immediately.
- Report which disbursements can be postponed.
- Report on overdue collections and possibility of collections of these accounts.

Petty Cash:
- Control petty cash fund.
- Require a voucher for each expenditure.
- Maintain monthly petty cash report and post to appropriate accounts.

Taxes:
- File employee and employer taxes on schedule.
- Maintain tax withholding records for each employee.
- Maintain appropriate accounts for employer taxes.

Bank Accounts:
- Reconcile each account monthly.

Records:
- Maintain records as required, including cash journal, general ledger, special accounts, inventory, depreciation, payroll, and others as required.

Financial Reports:
- Coordinate with accountant.
- Prepare periodic reports as required.
- Anticipate accountant's needs or request a list.
- Maintain all records in good order for review or audit.

Appendix

Job Description for Home Inspector

QUALIFICATIONS

1.   Knowledge of building construction gained through one or more of the following:

> ✓ Five years or more experience in home building, remodeling or renovation as a licensed contractor or working under a licensed contractor

> OR

> ✓ B.S. degree in architecture, engineering, or construction or satisfactory completion of a trade school program in any of these fields, plus two years of on-the-job experience.

> OR

> ✓ Demonstration of exceptional mechanical and construction knowledge obtained through self-education and at least two years of actual work experience in the construction or remodeling industry.

2.   The ability to relate to other people and discuss their concerns, problems, and fears concerning the home inspection process in a professional manner. The home inspector also should have the ability to discuss intelligently the real estate market, the national and local economy as related to home purchases, and technical aspects of home construction and components.

3.   The ability to complete a home inspection report in a clear and satisfactory manner and to discuss the report with the customer.

HOURS

The home inspector must be available to perform inspections and complete reports during business hours, 8 a.m. to 5 p.m., Monday through Friday.

The home inspector must be available for home inspections before and after business hours and on weekends, as required by the on-call rotation system of the company.

THE HOME INSPECTION

The home inspector will meet with the customer or the customer's agent at the appointed time at the inspection site. The inspector will briefly describe the inspection details to the customer, answer pertinent questions, and invite the customer to come along during the inspection.

The inspector will follow the checklist of items to be inspected and note on the checklist the condition, operation, life expectancy, and other observations as required by the checklist.

Exterior of the structure:
- Inspect siding and trim work.
- Inspect physical and safety conditions of porches, decks, steps, balconies, railings, carports, garages, driveways, walkways, patios.
- Inspect exposed areas of foundation walls, structural components, chimney foundations.
- Inspect grading and vegetation for conditions that could affect the structure or cause water problems.

Roof and attic:
- Inspect from ground level the roof type, materials, flashing, joints, soffits, fascias, skylights, and other roof accessories.
- If flat roof is safely accessible, inspect from roof itself.
- If attic space is accessible, inspect roof framing and underside of sheathing, note conditions due to poor ventilation, inspect attic insulation, and note visible evidence of water penetration.
- Inspect from ground level the rain gutters and downspouts.

Basement/crawl space:
- Inspect basement for obvious structural weakness, masonry cracks, water damage; note condition of finished surfaces, if present; operate windows and doors to ascertain availability of safe egress from basement.
- If crawl space has three feet or more of headroom, inspect this area for obvious structural damage, condition of heating ducts, plumbing lines, insulation, and other systems; note excessive dampness or other adverse conditions.

Interior:

- Inspect random number of windows and doors and hardware.
- Inspect stability and surface condition of walls, ceilings, floors, steps, stairways, balconies, railings.
- Inspect exterior condition and mountings of cabinets, counter tops.
- Inspect fireplace dampers, flue linings, fireboxes, and hearths.

Electrical system and appliances:

- Inspect masthead, conduits, distribution panels.
- Test and/or verify operation of representative number of switches, receptacles, light fixtures; grounding, polarity, bonding of receptacles on exterior and in proximity to plumbing fixtures; operation of ground fault circuit interrupter devices.
- Verify existence and condition of connected service grounding cables; compatibility and condition of main and branch circuits for overload protection in relation to size of conductors; safety of exposed wiring.
- Operate, where possible and practical, all kitchen exhaust systems, ranges, ovens, dishwashers, garbage disposers, and other kitchen appliances; laundry appliances; attic and other house fans.

Plumbing:

- Inspect all accessible and visible water and waste lines; unusual drainage conditions at fixtures; connections between stack and internal drainage, where visible; domestic hot water system; visible private water system equipment.
- Test for adequacy of water flow; functional drainage.
- Inspect and operate all plumbing fixtures and faucets, unless flow end is connected to an appliance; drain pumps and waste ejector pumps.
- Check for main water supply valve.

Heating and cooling:

- Operate heating and cooling systems using normal control devices; access and inspection panels, if provided.
- Inspect visually all controls and operating components of systems, conditions permitting; representative number of heat and cooling sources throughout house; supplementary heat components.
- Evaluate flue pipes, dampers, humidifiers, electrostatic air cleaners, motorized dampers, and other components.

The inspector is not required to:
- Perform a termite inspection.
- Walk on roofs where it could damage materials or be unsafe for the inspector.
- Remove snow or ice to gain visibility of roof or other areas.
- Enter spaces not readily accessible.
- Move furniture, carpeting, or personal goods that impede access or visibility.
- Inspect underground drainage pipes or internal rain gutter and down spout systems.
- Evaluate extent of insulation within exterior walls and finished ceilings.
- Evaluate adequacy of electrical ground cable.
- Activate electrical systems that are disconnected, or dismantle any electrical equipment or controls.
- Operate any components of plumbing, heating, or cooling systems that are turned off.
- Inspect any equipment not in readily accessible areas.

THE INSPECTION REPORT

The home inspector will make sufficient notes on the checklist or other form to render an accurate inspection report. Weather conditions and any unusual circumstances or events pertaining to the inspection also will be noted.

Notes must contain the date, time, and weather conditions of the inspection, names of all persons present at the inspection, and any pertinent requests from the client regarding the inspection or purpose of the report. It is not company policy to change terms of the pre-inspection agreement on-site; e.g., a partial inspection excluding electrical and mechanical equipment shall not be changed to include such equipment without prior clearance from the office.

The inspection report must be written or dictated immediately following the inspection (if not completed and delivered on-site) and delivered to the office the day of inspection or, in the event of late evening appointments, early the next day, with notes or tape submitted for transcribing and typing.

Appendix

COLLECTIONS

The home inspector will be responsible for collections on jobs inspected. This may be done by collecting a check on-site at the time of the inspection, getting credit card information, collecting cash, or getting billing information. With rare exceptions, company policy is to collect the inspection fee when the inspection is performed and before the report is delivered. If company authorizes billing of the client for later collection, the inspector is no longer responsible for this task. The inspector works as a subcontractor and is paid from funds actually collected, not from jobs billed.

Most Common Liability Claims

1. Roofs

Failure to detect an asphalt or fiberglass shingle roof that is near the end of its normal life is most common in darker roofs. In the lighter colors, wear is easier to tell.

In a metal roof that has been tarred, you cannot determine the condition of the metal under the roof. Failure to spell this out has resulted in claims.

Not knowing the difference between a Vermont and Buckingham slate or a Pennsylvania and a Bangor slate, has resulted in claims where customers felt they bought a lifetime roof when it was only a 30-50 year material and needed replacement.

2. Water

Most water problems in the basement or crawl space are not a result of a high water table but are due to a lack of roof and surface water control. However, any purchaser of a house with a water problem is going to be unhappy and in many cases they are going to take out that frustration on the home inspector. This problem is aggravated by the fact that most waterproofing contractors will recommend a french drain and sump pump at a cost of several thousand dollars, regardless of the cause of the problem.

3. Structure

Structural problems are a major liability issue, particularly in areas where there is marine clay or shifting soil. These claims are not as frequent as water or roof problems, but are usually of substantial financial magnitude.

4. Heat Exchangers

The entire heat exchanger cannot be seen in a visual inspection of gas or oil fired furnaces. Even with disclaimers, customers who move into a house and find they must immediately buy a new furnace are likely to seek recourse against the inspector.

5. Asbestos

Asbestos has been a major liability problem. If there is visible possible asbestos, particularly if it is friable, and the inspector does not call attention to the possibility that it is asbestos, there is heavy liability. There have been a number of law suits on this oversight.

6. Other Problems

Other problems are not as frequent, but you should still be on the lookout for them.

In the south, where air conditioning units run over 2,000 hours a year, these units are a high liability issue.

Hot water boilers tend to last longer and have fewer problems than forced air furnaces, but one problem that does crop up is circulating pumps that wear out.

According to ASHI standards, most inspectors disclaim or do not check well and septic systems, so there is very low liability in this area, but this disclaimer must be very clear. Common major plumbing problems include lead shower pans and ceramic tile in tub areas, particularly tile in mastic rather than mud.

Termite and insect infestation inspections are usually not part of the inspection, although in some parts of the country, inspectors do handle that. It is necessary, however, to look for termite structural damage, and this can be an area of liability if not checked carefully.

Although ASHI standards do not mandate inspection of kitchen appliances, most inspectors throughout the country do inspect them. These must be run through their cycle and checked, which creates some liability, although it does not have a major financial impact.

Sample Company Policy Manual

To ensure that employees and subcontractors clearly understand and can know what to expect from your company, a company policy manual is an important tool. This creates a perception of fair play and gives your people a "constitution" or set of written policies within which to operate. The manual states policies and procedures relating to employees, such as:

Vacation Policy: Many companies give employees a week of paid vacation after one year of full time employment, and two weeks after two years. This should be clearly specified in your manual.

Paid Holidays: Companies often pay employees for several holidays per year, such as Christmas, New Years Day, Memorial Day, Labor Day, and one or more additional national or religious holidays. Many companies pay for the holiday only if the employee works the business day before and after the holiday.

Payday: Many companies pay on a weekly basis, on Friday for all work completed through the previous Tuesday or Wednesday. This allows time for the accounting department to calculate paychecks based on time cards or inspection collections. It is important that all new employees or subcontractors understand this policy.

Working Hours: Companies should specify the time work starts, the number and duration of all breaks including lunch, and closing time.

Vehicles: If the company pays a mileage allowance for vehicles owned by employees and used for company business, the policy should be stated. The rate of pay per mile, etc., may require keeping a log, displaying magnetic signs during company business, keeping fuel and maintenance bills, and other items.

Employees tools: Many companies will set up a standard list of tools each home inspector should supply. These may include a cassette tape recorder, on-site tools, home answering device, etc. The company may provide certain tools needed for the daily conduct of business, and if so, this should be stated at the outset of the business relationship and included in the manual.

Appendix

Use of Company Equipment: Many companies require equipment to be checked out when it is taken to the field. The responsibility for care and condition of the returned equipment should be stated in this manual.

Alcohol, food, and smoking on the job: Most companies expect professional performance and behavior from home inspectors as well as other office employees. This will mean no use of alcohol on the job, no accepting food, coffee, etc., from clients, no smoking on the site of an inspection. While office policy may be more lenient towards coffee drinking and smoking on the premises, these expectations should be spelled out in the manual.

Overtime: Without express permission of the office manager or job superintendent, overtime is not normally allowed. This does not apply to subcontractors but does apply to office personnel, and should be stated.

Moonlighting: It is common policy for companies not to allow moonlighting. This may apply to full time employees as well as subcontract inspectors, and should be explained in the manual.

Sick Leave: Sick leave and annual leave are not allowed by some companies, allowed by others. If policy states the terms of leave and how it is accrued, there will be no reason for any misunderstanding down the line.

Call-In Time: Field people may be required to call in at specific times of day for inspection and other job assignments. This will be understood in practice, but should be spelled out in the policy manual as well.

Miscellaneous: Additional policies may apply to home inspectors or professional level people and not to clerical level, or the reverse. In many cases these specific policies will be spelled out in a separate sheet or manual booklet.

Inspection on an Old House

A HomeTech inspector knew in advance the intended use owners had for this report on an old house in Fredericksburg, Va. They wanted information that would give them a strategy and perhaps lead them into a go/no-go decision on the purchase of an antebellum house which might need major structural work. In addition, they would need to know something about the extent of work necessary to upgrade a dilapidated structure to modern comfort and efficiency standards.

This kind of inspection presents no more liability danger to the company than any other, but due to the extensive work which will be needed, disclaimers and cautions must be clearly and repeatedly stated. It is important to give estimated ranges for renovation and repair items. Because complete replacement or repair of whole systems are anticipated, the best approach is the systems approach. (Another true-to-life letter from the HomeTech files.)

Dear Mr. and Mrs. _____:

Here are the results of our inspection conducted on the property at _____, Fredericksburg, Virginia. This inspection was conducted using standard, visual techniques. There are no warranties of any type expressed or implied.

Structural: This is a masonry and frame row house dwelling built in stages, dating from approximately the 1850s.

The building shows signs of considerable movement, but we believe it can be stabilized with some shoring up. We remind you that this structure was not built using the same design criteria as today's homes, and loading characteristics are different. A modern house is built to support about 40 pounds per square foot on the floor, where this kind of house might hold 20-25 pounds per foot. This means some kinds of modern furniture, such as waterbeds, could not be used without reinforcing the house floor structure.

We recommend a complete exterminator's treatment and contract, along with an insect report. HomeTech is not equipped to do insect infestation reports, which require a licensed exterminator.

Appendix

The masonry finish on the outside needs complete repointing for cosmetic appeal and structural soundness. The old brick are weathered and mortar has eroded over the years.

We believe this house will require complete gutting and renovation of all systems, structural reinforcement, and all cosmetics.

Electrical: There is essentially no useable electric service in the building. We recommend a complete rewiring. Depending on your choice of heating system, we suggest a 150 amp or 200 amp service. If you use electric heat pump or central air conditioning, in addition to other major electric appliances (water heater, range, dryer, etc.), the 200 amp system is recommended.

Heat/Ventilation/Air Conditioning: There are no useable HVAC systems in the building presently. We suggest you investigate using a high efficiency system such as a heat pump.

Plumbing: Antiquated plumbing should be removed and a complete new plumbing supply and drain/waste/vent system installed. This is consistent with the gutting/renovation mentioned earlier.

Cellar: The exterminator was on the premises when I arrived and had discharged a large quantity of insecticide into the cellar area. Fumes made it impossible to observe this portion of the structure in any detail. However, I discussed the structural condition with the exterminator and did not discover major structural problems. With the building in basically its original condition as evident from other accessible areas, we are confident there will be some shoring up necessary in the cellar.

Kitchen: There is no useable kitchen equipment in the building at this time.

General Interior: Interior walls and ceilings are plaster over wood lath in need of considerable cosmetic attention. Flooring is basically random width heart pine throughout the two and one-half stories. While the flooring is salvageable, it will be a considerable undertaking.

Windows are single glazed, double hung wooden sash units in poor condition. We recommend these windows be replaced; however, since the Historic Fredericksburg requirements are quite strict, you should contact the appropriate officials before undertaking any work on the exterior.

Insulation in the building is essentially non-existent. We recommend adding R-30 insulation with proper ventilation, or as close to that rating as can be fitted in, in attic spaces wherever possible. Exterior walls are solid masonry and insulation will be very difficult to add without rebuilding the interior trim, which has considerable aesthetic appeal. Since heat loss through the walls is less important than the ceiling, we suggest you pay most attention to windows and ceiling insulation. Floors between living space and the cellar should also be insulated to R-19, and to meet modern energy standards, the walls facing the outside elements (not party walls between house units) could be furred and receive foam board insulation with new sheetrock interior finish walls. This furring treatment will also provide space for electrical conduit to enable you to have wiring and receptacle boxes on outside walls.

Fireplaces were not lined with firebrick, equipped with operable dampers, or attached to terra-cotta lined flues. All should be components of safe, modern units. Fireplaces throughout the building should be rebuilt if they are to be used.

General Exterior: It was snowing excessively at the time of inspection and the roof shingles, apparently slate, were all but obscured from view. If slate, they are one of two types, probably Pennsylvania.

There are slate roofs and there are slate roofs. Vermont slate roofs can last for 75 or 100 years or more. Although they sometimes have problems with rusted nails or loose shingles, they are durable. Nails or individual slates may require periodic replacement.

Pennsylvania slates, by contrast, have a normal lifetime of 30 to 50 years, at which time they shale and spall. Once this starts happening, you have an expensive and ongoing maintenance problem. Many home buyers find the best decision at this point is to take off the slate and install solid sheathing and composition shingles to minimize expense, which would likely be the least expensive avenue in this instance.

Roof flashing is in poor condition, showing signs of leaks in several places along the upper ceilings. Gutters and downspouts need to be replaced.

Repointing masonry is needed in several places on the exterior, possibly most of the brickwork needs this treatment.

Appendix

The following items may be reasonably expected during the renovation of this building.

1.	General expenses	$5,000-7,000
2.	Complete rewiring, with 150 amp or 200 amp main breaker panel	$3,000-5,000
3.	Installation of ducting and HVAC system, either gas or heat pump	$6,000-8,000
4.	Install new kitchen	$8,000-15,000
5.	Install bathrooms (each bathroom)	$5,000-8,000
6.	Plumbing rough-in for two baths, kitchen	$2,000-3,000
7.	Install new roof, including sheathing	$3,000-5,000
8.	Point up masonry	$ 800- 1,100
9.	Cosmetic interior repairs	$5,000-7,000
10.	Insulation	$1,200-2,300
11.	Rebuilding fireplaces (each opening)	$2,000-2,500
12.	Rebuild or replace windows	$6,000-10,000

We would caution you that furnaces and all combustion appliances must have individual flues. Fireplaces must each have individual flues. These costs should be accounted for in planning for a new heating system.

We believe it would be reasonable to anticipate spending well over $50,000 renovating this property.

Sincerely,

REPORT ANALYSIS

A major renovation job can present the owner with nasty surprises at many points along the way. Because inspectors are not engineers, architects, designers, or contractors, the report must be clear but have a "back door". Clarity is the mark of a professional job. It reassures the client. Systems must be addressed completely, and the report must communicate the likelihood of alternative solutions and choices which are up to the owner.

Here's the back door. By careful analysis of each system, coupled with a comfortable cost range, you give the clients a workable tool to be part of their decision and strategy. The cost ranges must stand up to the reality of the local marketplace, but rather than stating a range so wide the client may think you don't know your business, you can make a brief general statement. To say that it would be "reasonable to anticipate spending well over . . ." challenges the clients' creativity. It lets them know the sky's the limit and you're not holding a safety net on any choice they may make.

INDEX

INDEX

INDEX